A Colour Atlas of
Medical Entomology

CHAPMAN & HALL MEDICAL ATLAS SERIES

Chapman & Hall's new series of highly illustrated books covers a broad spectrum of topics in clinical medicine and surgery. Each title is unique in that it deals with a specific subject in an authoritative and comprehensive manner.

All titles in the series are up to date and feature substantial amounts of top quality illustrative material, combining colour and black-and-white photographs and often specially developed line artwork.

The amount of supporting text varies: where the illustrations are backed-up by large amounts of integrated text the volume has been called 'A text and atlas' to indicate that it can be used not only as a high quality colour reference source but also as a textbook.

Slide Atlases are also available for some of the titles in the series.

1. **A Colour Atlas of Endovascular Surgery**
   R.A. White and G.H. White
   Also available:
   A Slide Atlas of Endovascular Surgery

2. **A Colour Atlas of Heart Disease**
   G.C. Sutton and K.M. Fox

3. **A Colour Atlas of Breast Histopathology**
   M. Trojani

4. **A Text and Atlas of Strabismus Surgery**
   R. Richards
   Also available:
   A Slide Atlas of Strabismus Surgery

5. **A Text and Atlas of Integrated Colposcopy**
   M.C. Anderson, J.A. Jordon, A.R. Morse and F. Sharp
   Also available:
   A Slide Atlas of Colposcopy

6. **A Text and Atlas of Liver Ultrasound**
   H. Bismuth, F. Kunstlinger and D. Castaing

7. **A Colour Atlas of Nuclear Cardiology**
   M.L. Goris and J. Bretille

8. **A Colour Atlas of Diseases of the Vulva**
   C.M. Ridley, J.D. Oriel and A.J. Robinson

9. **A Colour Atlas of Burn Injuries**
   J.A. Clarke

10. **A Colour Atlas of Medical Entomology**
    N.R.H. Burgess and G.O. Cowan

11. **A Text and Atlas of Arterial Imaging**
    D.M. Cavaye and R.A. White

12. **A Colour Atlas of Respiratory Infections**
    J.T. Macfarlane, R.G. Finch and R.E. Cotton

13. **A Text and Atlas of Paediatric Orofacial Medicine and Pathology**
    R.K. Hall

In preparation

A Text and Atlas of Clinical Retinopathies
P.M. Dodson, E.E. Kritzinger and D.G. Beevers

A Colour Atlas of Retinovascular Disease
S.T.D. Roxburgh, W.M. Haining and E. Rosen

A Colour Atlas of Forensic Medicine
J.K. Mason and A. Usher

A Colour Atlas of Neonatal Pathology
D. de Sa

A Text and Atlas of Breast Cytodiagnosis
P.A. Trott

A Colour Atlas of Medical Entomology

N.R.H. Burgess
Defence Adviser in Entomology and Senior Lecturer in
Entomology,
Royal Army Medical College, London, UK;
Adjunct Professor of Preventive Medicine (Entomology),
Uniformed Services University of Health Sciences,
Washington, DC, USA;
Honorary Senior Lecturer,
Imperial College of Science, Technology and Medicine,
University of London, London, UK

G.O. Cowan
Professor of Military Medicine,
Royal Army Medical College
and Royal College of Physicians,
London, UK

SPRINGER-SCIENCE+BUSINESS MEDIA, B.V.

First edition 1993

© 1993 Springer Science+Business Media Dordrecht
Originally published by Chapman & Hall in 1993

Typeset in 10/12 Palatino by Keyset Composition, Colchester, Essex

ISBN 978-94-010-4676-3   ISBN 978-94-011-1548-3 (eBook)
DOI 10.1007/978-94-011-1548-3

*For Sue, Miranda and Crispin (who feature
in several photographs), and for Bea.*

Contents

Acknowledgements

We are most grateful to the following colleagues who kindly provided figures for use in this Colour Atlas:

Dr M. Anderson, University of Birmingham (Figs 7.3, 8.5, 8.6, 8.7)

British Museum (Natural History) (Fig. 16.3b)

Dr D. A. Burns, Leicester Royal Infirmary (Figs 9.5, 9.15–9.17, 10.17, 11.11b, 14.3, 15.1, 15.2)

Canadian Forces Manual of Pest Control, 4th edn (1981), Department of National Defence, Canada (Figs 1.5, 2.2, 4.1, 7.23, 10.20, 11.1, 14.6)

Charing Cross and Westminster Medical School, London (Figs 14.11a, 14.11b, 14.15)

K. N. Chetwyn (Figs 2.13, 2.17, 2.22, 16.11)

Dr A. C. Chu, Royal Postgraduate Medical School, London (Figs 2.31, 10.18, 11.6, 11.10)

Dr I. P. Crawford (Fig. 16.19)

Dr D. France, Crawley Hospital (Figs 10.16, 15.10a, 15.10b)

Department of Medical Illustration, Manchester Royal Infirmary (Figs 11.12a, 11.12b)

Jeffrey, H. C. and Leach, R. M. (1975) *Atlas of Medical Helminthology and Protozoology*, 2nd edn, Churchill Livingstone, London (Fig. 2.38)

Dr A. M. Jordan, Tsetse Research Laboratory, Bristol University (Fig. 8.2)

J. R. Latimer (Figs 9.9a, 9.11, 9.13b)

Marshall, J. F. (1966) *The British Mosquitoes*, Johnson Reprint Company Limited, London (Figs 2.1, 2.14)

Nash, T. A. M. (1969) *Africa's Bane*, Collins, London (Fig. 8.11)

Oldroyd, H. *British Bloodsucking Flies* (BMNH) (Figs 6.7, 6.8)

Royal Army Medical College (Figs 3.15, 3.16a, 3.16b, 8.1, 15.10c, 15.10d, 16.10, 16.20)

S. McDermott (Figs 2.6, 2.21, 2.35, 4.5, 7.4, 7.7–7.12, 7.19, 8.4, 9.7, 9.19, 11.3, 12.2, 16.15)

Sandoz Ltd, Argo Division, Switzerland (Figs 2.7, 2.20, 2.24a, 2.29, 17.15)

Dr S. Selwyn, Westminster Hospital (Fig. 15.5)

Professor D. Taplin and Dr Terri Meinking, University of Miami, USA (Figs 7.21c, 10.6, 10.9, 10.12, 15.7)

A. M. Walker (Figs 3.7, 11.7)

Walter Reed Army Institute of Research, Washington, DC, USA (Figs 2.36b, 2.37a, 3.14a, 3.14b, 3.14c, 3.14d)

N. G. Williams (Figs 8.10, 14.13)

World Health Organization (Figs 2.9, 2.10, 10.7, 13.2, 13.6)

The authors would like to acknowledge the assistance of the Department of Military Entomology and the Department of Medical Illustration, Royal Army Medical College, in the preparation of this work.

Over the last 30 years copies of many teaching transparencies have changed hands, and every effort is always made by authors to give due acknowledgement. If discrepancies have occurred, please advise the authors and accept our sincere apologies.

Preface

Insects and their allies affect humans in a wide variety of ways. As well as being highly beneficial as pollinators of fruit and flowers, many are attractively coloured and some even sound pleasant. But they also cause considerable detrimental effects, destroying crops and damaging food, property, and livestock. Most significant is their role in transmitting disease to humans and causing debilitating conditions and extreme discomfort.

This Atlas, which is illustrated with many of the authors' own photographs, will enable the reader to identify the cause of these problems and, with a knowledge of the creatures' habits, habitat and life history, to prevent further attack or infestation.

The authors are indebted to their colleagues who have assisted in the production of the Atlas, and to those who have allowed their photographs to be used.

N. R. H. Burgess
G. O. Cowan
London

1. Introduction

Members of the phylum Arthropoda are the most numerous and widely distributed of all animal groups. They may be found in every part of the world and in every type of environment. Many, particularly those within the classes Insecta and Arachnida, live in close association with humans; others, while primarily parasites of animals, will readily attack or feed upon humans and some may be specifically adapted as human parasites. They may be of medical significance simply because of their physical attack and blood-sucking habits, or they may be of considerable importance as distributors of organisms that cause disease. These arthropod disease vectors may transfer pathogenic organisms in a purely mechanical manner from infected material to human food, or, more significantly, they may act as developmental vectors, incubating the disease organisms in their bodies before passing them on to uninfected hosts. Several arthropods, for example beetles, moths, cockroaches and mites, may be of considerable public health and economic importance, causing significant damage and nuisance by infesting and feeding on stored food and other commodities and infesting domestic situations.

All arthropods, while varying considerably in size and shape, have certain features in common (Fig. 1.1). They are all bilaterally symmetrical and metamerically segmented. They have a hard chitinous exoskeleton, sometimes sclerotized or calcified, inside which is a hollow blood cavity containing a clear fluid (haemolymph), a dorsal tubular heart, alimentary tract and a central nervous system of two longitudinal nerve trunks fused segmentally to form ganglia. Respiration may be achieved by a variety of methods: via gills, lung-hooks, gaseous exchange through the cuticle or by means of spiracles. All arthropods have jointed appendages, which may

Fig. 1.1 A typical arthropod, the centipede, with exoskeleton and jointed paired appendages.

take the form of legs, antennae, mouthparts or cerci. The sexes are always separate. Some of the main classes are listed in Table 1.1.

THE PHYLUM ARTHROPODA AND ITS CONSTITUENT CLASSES

Classification

Every living organism is named in Latin according to the genus (first word) and the species (second word). For example, the housefly, *Musca domestica*, and the malarial vector *Anopheles gambiae* are classified as shown in Table 1.2.

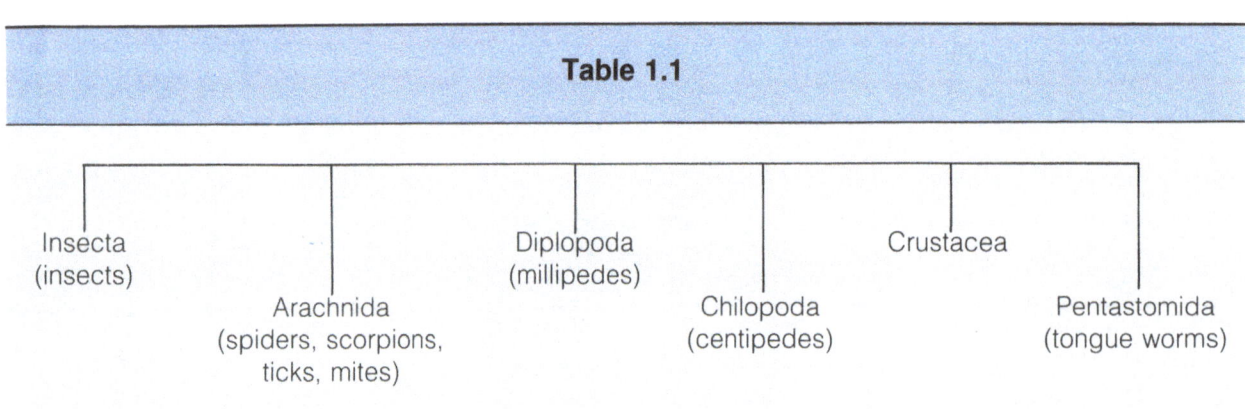

Table 1.1

Insecta (insects)

Arachnida (spiders, scorpions, ticks, mites)

Diplopoda (millipedes)

Chilopoda (centipedes)

Crustacea

Pentastomida (tongue worms)

The classification of living organisms is essentially a manufactured system set up for convenience. Thus, as more insects are discovered and described each year, the gaps in groups will be filled and the variations and similarities between species will become more apparent. Furthermore, insects continue to evolve. For these reasons, or even because of the

according to diet, and usually paired eyes, which may be compound or simple. The thorax supports three pairs of legs and typically one or two pairs of wings (Fig.1.3). The abdomen is usually without appendages except for the presence of genitalia on the posterior segments.

The exoskeleton protects the viscera and prevents

Table 1.2		
Phylum	Arthropoda	Arthropoda
Class	Insecta	Insecta
Subclass	Pterygota	Pterygota
Division	Endopterygota	Endopterygota
Order	Diptera	Diptera
Suborder	Athericera	Nematocera
*Series	Schizophora	—
*Section	Calpteratae	—
*Superfamily	—	—
Family	Muscidae	Culicadae
Subfamily	Muscinae	Anophelinae
Genus	*Musca*	*Anopheles*
*Subgenus	—	—
Species	*domestica*	*gambiae*
*Subspecies	—	—
*Variety	—	—

whims and inclinations of taxonomists, the system of classification is continually changing, hence subfamilies may be given family status, or one genus may be divided into several new ones. The classification adopted in this atlas is given for guidance. The description of the insect, whether it be of a family or subfamily, a genus or species, does not change.

ANATOMY

The class Insecta forms the largest of all animal classes, with about 1 million described species. However, only a very small proportion are of medical or public health significance. In the adult stage, insects are characterized by the division of the body into head, thorax and abdomen (Fig. 1.2). The head always bears one pair of sensory antennae, three pairs of mouthparts, which are modified

water loss. Moreover, the internal organs (Fig. 1.4) are supported by body fat, which also acts as a food store for the insect in unfavourable conditions.

The gut is divided into three sections: the foregut, mid-gut and hind-gut. The fore-gut projects backwards from the mouth towards the thorax and is flanked by salivary glands, which are of particular importance in the transmission of disease. Behind the salivary glands, the fore-gut expands to form the crop or proventriculus. The crop is lined with nodules or spikes of chitin, which grind the food as an aid to digestion. In the mid-gut further digestion and absorption take place, and the posterior of the mid-gut often possesses several blunt tubules known as Malpighian tubules or hepatic caecae, which serve the same function as the liver in higher orders of animals. Nitrogenous waste is excreted at the narrowing of the tubules, which perform a similar function to the kidney. The Malpighian tubules join the gut at the intersection of the mid and hind-gut. The function of the hind-gut is to reabsorb water and, in some cases, salts. The hind-

Fig. 1.2 Gatekeeper butterfly, *Pyronia tythonus*. A typical insect with the head bearing a pair of antennae, mouthparts and compound eyes, the thorax with three pairs of legs and two pair of wings, and the abdomen without appendages.

Fig. 1.3 Horsefly, *Hybomitra distinguenda*. A typical insect, but with only one pair of wings (order Diptera).

gut terminates at the anus, which is flanked by two rectal glands.

The circulatory system is open, unlike that of humans (which is a closed system). Haemolymph or 'blood' is a colourless fluid in insects and does not contain haemoglobin. One exception is the larva of the midge *Chironomus*, which lives in water with a low supply of dissolved oxygen. Its haemolymph contains small amounts of haemoglobin as an aid to respiration, hence its red colour. Haemolymph is pumped through the body by the dorsal tubular heart (which is a long segmented tube) towards the head, where it flows into sinuses or cavities and percolates through the entire body, bathing the organs and distributing dissolved nutrients.

The nervous system is composed of a double ventral nerve cord with a series of ganglia; these are clusters of nerve tissue that form a node in each body segment. The ganglia increase in size towards the head where the nerve cord divides to encompass the fore-gut, with a large ganglion above and below the oesophagus. These two ganglia can be termed the 'brain' of the insect. Each ganglion has nerve branches leading to it from all parts of the body to receive and transmit stimuli. The body ganglia are often fused, particularly in the abdomen.

Insects do not possess lungs, and respiration takes place in the tracheal system. Each abdominal segment typically possesses two spiracles situated laterally. The thorax has four larger spiracles, two placed anteriorly and two posteriorly. The spiracle is a pore leading to a corrugated tube (trachea) that divides and redivides into a complex system of branches (tracheoles), permeating throughout the insect body,

Fig. 1.4 Internal anatomy of a typical insect (cockroach). The internal organs are supported by body fat.

and provides a large surface area for gaseous exchange. Oxygen entering the trachea is dissolved on the moist surface of the tracheal lining and absorbed directly. Respiration is a continuous process and no tidal breathing occurs. The number of spiracles in the abdomen may be reduced in some insects.

Separate sexes always occur in insects. The gonads or sex organs are internal and consist of testes in the male and ovaries in the female; they can be seen on either side of the hind-gut of mature specimens. Female insects possess a sperm sac or spermatheca, which stores live sperm after mating. With this, the female insect can control the amount of sperm released to fertilize future eggs and thus only needs to mate once in her lifetime. Male insects possess a penis or aedegus and copulation occurs during mating.

All arthropods possess an exoskeleton composed of chitin, a rigid waxy substance. The dorsal and ventral sections, the tergum and sternum respectively, are heavily chitinized. The lateral section, joining the tergum and sternum (the pleuron) is less heavily chitinized and thus more flexible.

The class Insecta is divided into two subclasses, the Apterygota and the Pterygota.

Apterygota

Members of the Apterygota (a = without; pteron = wing, in Greek) are believed to be the oldest form of insect. They never develop wings and typically have one or more pairs of well-developed appendages on the abdomen in addition to the genitalia. Once hatched from the egg, the apterygote insect increases in size but there is very little change in shape (metamorphosis). Apterygote insects include the springtails (Collembola) and bristletails (Thysanura), for example the silverfish and firebrat (Fig. 1.5). The latter may infest kitchens and other warm and humid places, where they feed on starchy scraps at night. Occasionally, numbers of firebrats need to be controlled but they cause little harm.

Pterygota

The pterygote insects are divided into two quite separate groups or divisions, the Exopterygota and the Endopterygota, and are differentiated by their method of development from egg to adult.

Exopterygota

These insects tend to lay comparatively few large eggs from which six-legged nymphs hatch. These resemble the parents in almost every way except for

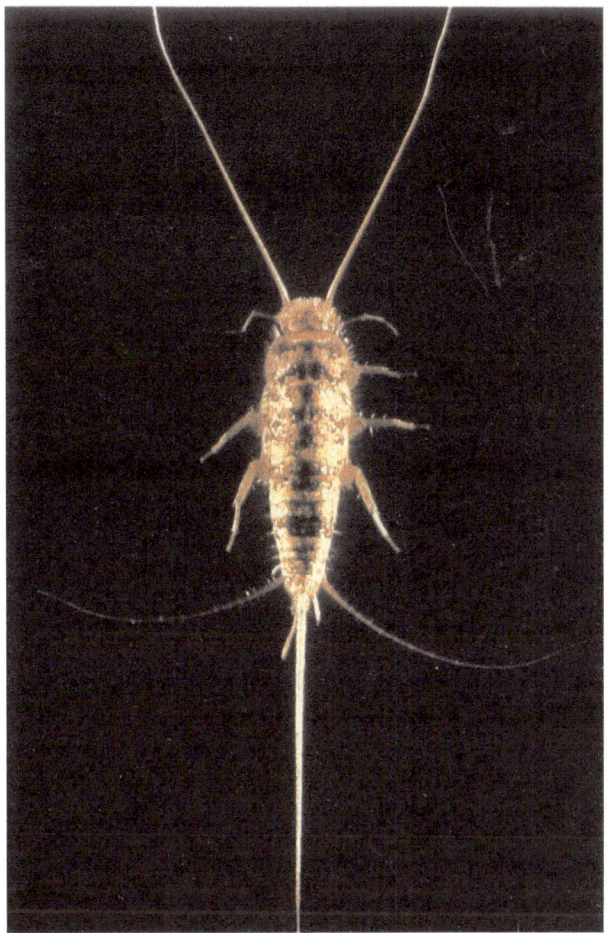

Fig. 1.5 A typical apterygote insect, the firebrat (Thysanura).

their smaller size, lack of flying wings and sexual maturity. Typically, the wings are visible externally as wing buds, which increase in size with each moult, hence exo (= outside) pterygota (= wings). (The term larva is sometimes used in this context to refer to the first-stage nymph.) The nymphal stages (instars), of which there may be up to ten or more depending on the species of insect, feed and increase in size by casting their skins (moulting or ecdysis) periodically.

Endopterygota

These insects tend to lay many eggs, usually on the larval food supply. From each of these eggs, a larva is hatched with little or no resemblance to the adult form. The larval stages, of which there may be several, eat voraciously and moult to increase in size, until they have accumulated sufficient food material to undergo a non-feeding and usually sessile stage of development known as the pupa. The sexually mature, winged adult form develops inside the pupal case and eventually emerges. The wings develop inside the body cavity within the pupal case and are evaginated on emergence, hence endo (=inside) pterygota (= wings).

The life cycle of an exopterygote insect may be described as incomplete metamorphosis, whereas that of an endopterygote insect has a complete metamorphosis.

SIGNIFICANCE TO HUMANS

Insects and related arthropods affect humans adversely in many different ways. They may feed on foodstuffs that are still growing or during processing or storage, and they may damage and parasitize domestic animals, lowering their condition; however, most significantly they may attack humans themselves, causing irritation, discomfort and inconvenience, and will often transmit disease-causing organisms.

In temperate regions, faster and more comprehensive air travel has significantly increased the danger of imported disease, often into areas in which it may have been endemic in the past and where the insect vector still exists. In addition, the increased incidence and greater awareness of the role played by arthropods in the more domestic scene has emphasized the need to control public health pests such as cockroaches, fleas, lice and mosquitoes. In the tropics, arthropod-borne diseases have devastated human populations in the past and continue to be an important cause of severe morbidity and mortality. Many diseases are re-emerging in developing countries due to economic problems and the resistance of vectors to insecticides and of pathogenic organisms to therapeutic drugs. The need to control arthropods, both from the public health and the medical point of view, has never been greater, and the task is becoming increasingly more difficult.

Worldwide, the situation is complicated by the attitudes of authorities and minority lobbies and the essential legislation imposed on the registration and use of pesticides, which is further handicapped by the vast financial outlay that this implies.

DEVELOPMENTAL TRANSMISSION OF DISEASE

The most important form of attack by arthropods is that involving the blood-sucking habits of ectoparasites, such as mosquitoes, bedbugs and fleas. These may simply be the cause of irritation and discomfort, sometimes resulting in secondary infection of the bite. More significantly, blood-sucking insects may transmit the organisms causing several debilitating and sometimes fatal diseases such as malaria, plague, typhus and yellow fever.

Typically, when an insect bites, a small amount of anti-coagulant saliva is injected into the wound to prevent the blood from clotting. It is the host's antibody reaction to this saliva that results in the swelling, redness and irritation at the site of the attack. It is a fact that some people will be bitten more regularly than others; the reason for this is unknown but it may be due to skin colour and texture, body temperature and content of perspiration. It is also true that some people react more apparently and more rapidly than others; this reflects the state of immunity and amount of exposure to bites from that particular insect in the person concerned. Secondary infection, which can be severe, is caused by bacteria entering a bite puncture or an area that has been traumatized by scratching.

In temperate regions of the world, the most common insect bites are those of mosquitoes and other blood-sucking flies, bedbugs, fleas and lice. Fortunately, these insects, while commonly transmitting disease in the tropics, do not often do so in temperate regions. Nevertheless, malaria, carried by the *Anopheles* mosquito, was endemic in many cooler parts of the world until recent times, and several locally transmitted cases occur every year in people who have never been to an endemic area. In addition, the number of cases of malaria imported into temperate countries, where the erstwhile vectors are still present, is considerable. Some mosquito-borne viruses and tick-borne bacteria are the cause of sporadic disease outbreaks in temperate regions including the British Isles. In the tropics and subtropics, the role of insects and other arthropods in the transmission of disease is dramatic.

PHYSICAL ATTACK BY INSECTS AND OTHER ARTHROPODS

Many arthropods will attack humans for a variety of other reasons. Sometimes it will be an unintended contact, often causing an allergic response as, for example, asthma from the house-dust mite, urticaria from certain moths and caterpillars, or dermatitis in food handlers caused by pests found in stored products. Bees, wasps, ants and scorpions may sting to protect themselves when provoked, and spiders and centipedes may bite. Flies will lay their eggs or larvae on open wounds or sores and the maggots will feed on necrosing tissue, producing the condition myiasis. The larvae of a few species, such as the tumbu fly and bot fly, can only develop on living tissue.

It must also be appreciated that entomophobia, or the fear of insects, is a very real pathological condition, and that delusory parasitosis is common.

MECHANICAL TRANSMISSION OF DISEASE

Insects that habitually feed on potentially infected matter and are closely associated with humans may carry pathogenic organisms (particularly those causing enteric disease) from one to the other.

Cockroaches, although more common in the warm tropics from where they originate, are frequently found in temperate areas, where they infest kitchens, restaurants and store-rooms. They will feed on refuse, human faeces and other potentially infected matter, and a wide range of pathogenic organisms have been isolated from these insects in domestic situations. Cockroaches frequenting such areas will walk over food intended for human consumption, on which they may defaecate and vomit and deposit

disease organisms. The adult female fly lays her eggs on decomposing organic matter in which the maggots will develop. At the same time both male and female flies will feed on decomposing matter, taking organisms into the gut, mouthparts and onto the body surfaces. Flies also commonly feed on human foodstuffs onto which they regurgitate digestive fluid containing organisms from a previous meal. They will also defaecate and deposit particles adhering to the feet and body hairs.

Thus, as well as causing a considerable nuisance by their presence (often in large numbers), cockroaches and flies may act as mechanical carriers of disease. This is a non-specific vectorial role since the diseases are also spread in a wide variety of other ways: in water, contaminated food and by human carriers. The insect role is contributory and circumstantial. The effects of arthropods on humans can be very varied. An appreciation of the appearance, lifestyle and habits of these creatures is an invaluable aid to an understanding and identification of many medical conditions. Greater knowledge will make treatment and cure more easily attainable and will render the prevention of recurrence more realistic by controlling the source of the problem.

2. Mosquitoes

INTRODUCTION AND DESCRIPTION

Mosquitoes belong to the order Diptera, a group of insects that only have one pair of wings, located on the mesothorax. The hind pair is reduced to small, drumstick-like organs, the halteres. Diptera can only take fluid food, sometimes in the form of blood. All species go through a complete metamorphosis in the life cycle.

Mosquitoes make up the family Culicidae and are small, midge-like flies, 5 to 15 mm in length and with a similar wingspan, a long narrow body and wings, and long delicate legs (Fig. 2.1). The most characteristic feature is the long, forwardly-projecting proboscis, which contains the mouthparts.

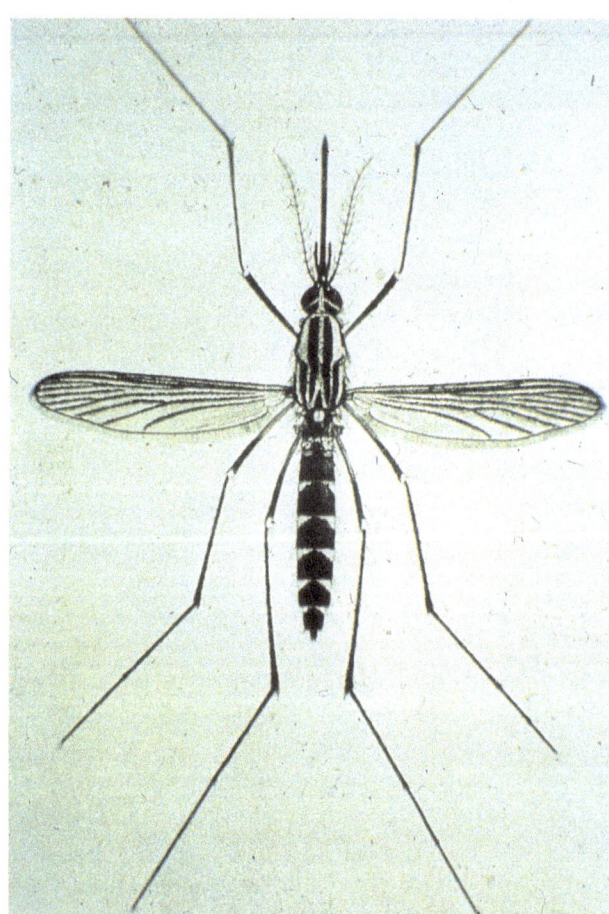

Fig. 2.1 Typical mosquito.

Some 3200 species of mosquito occur worldwide, and are classified into three subfamilies within the family Culicidae. Most species of medical importance are within the subfamilies Anophelinae (of which the most important genus is *Anopheles*), and

Culicinae (*Aedes*, *Culex* and *Mansonia*).

There are many species of non-biting gnats and midges (Fig. 2.2), which resemble mosquitoes in several features, but which do not have the long, forwardly-projecting proboscis. All mosquito species have a characteristic wing venation (Fig. 2.3), the third vein being short and simple, and lying between two forked veins. The wings are covered and fringed with scales.

Both male and female mosquitoes feed on plant fluids and nectar. However, the female typically requires a blood meal from a warm-blooded animal before a viable batch of eggs can be laid, and only the female is capable of sucking blood. The internal male mouthparts are short and only extend about a quarter of the length of the proboscis. In contrast, the female has long, needle-like mouthparts (Fig. 2.4), which are capable of piercing animal tissue.

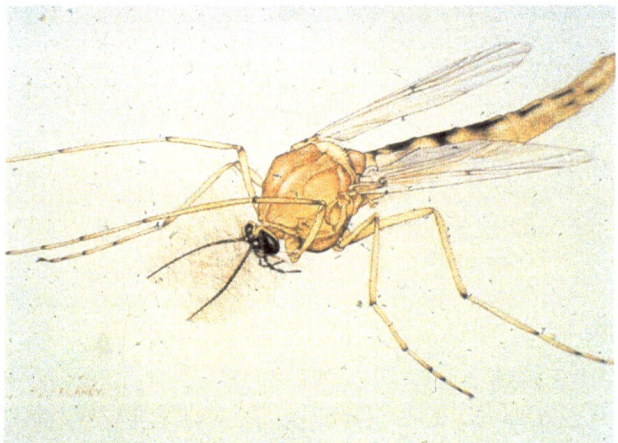

Fig. 2.2 Non-biting gnat. Note the absence of the long forwardly projecting proboscis.

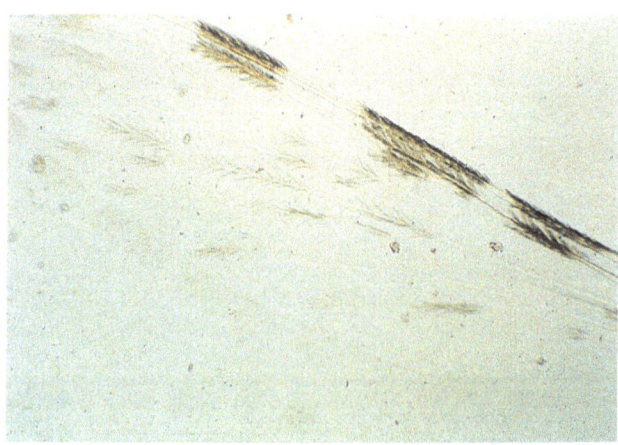

Fig. 2.3 Wing venation. The third vein is short and simple and lies between two forked veins.

Fig. 2.4 Mouthparts of the female mosquito. These are long and needle-like.

(a)

Male mosquitoes have a pair of long bushy (plumose) antennae (Fig. 2.5a), whereas the antennae of the female are sparsely haired or pilose (Fig. 2.5b). Both sexes have one pair of compound eyes.

GENERALIZED LIFE CYCLE AND BREEDING SITES

During the life cycle, mosquitoes undergo a complete metamorphosis, passing through the stages of egg, larva and pupa before becoming adult (Fig. 2.6). The immature stages are always associated with free water, which may occur in a wide range of locations. The mosquito egg hatches into a minute, worm-like larva (Fig. 2.7), which feeds on microorganisms in the water or on the water surface using paired mouth brushes on the head. Vision is rudimentary but larvae react rapidly to changes in light intensity, moving actively with a wriggling or darting motion through water.

The bulky, thoracic part of the larva often has long bristles or hairs, which assist in achieving balance. The abdomen has ten segments; the tenth segment is at an angle to the ventral surface and has anal papillae. On the dorsal surface of the ninth segment are the paired spiracles, sometimes on the end of an extended siphon, through which the larva obtains oxygen from the air/water interface. The larva passes through four stages or instars before moulting to the pupal stage.

The pupa (Fig. 2.8) is comma-shaped, the head and thorax having fused to form a cephalothorax, with the abdomen hanging down from it. The pupal stage is actively mobile, using a pair of paddles

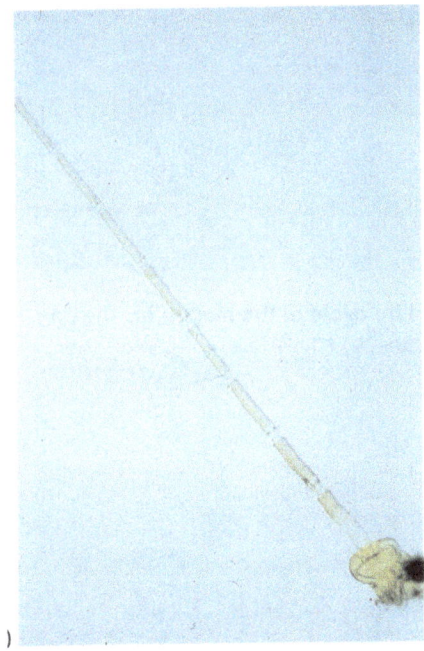

(b)

Fig. 2.5 Mosquito antennae. (a) Male. (b) Female.

located on the hind end of the abdomen to progress in a tumbling motion through the water. It does not feed but comes to the air/water interface to obtain oxygen through a pair of dorsal trumpets on the cephalothorax. The adult mosquito can be seen developing through the pupal skin.

When the adult mosquito is fully developed, the pupa comes to the surface and splits across the dorsum, the adult emerging to stand on the water

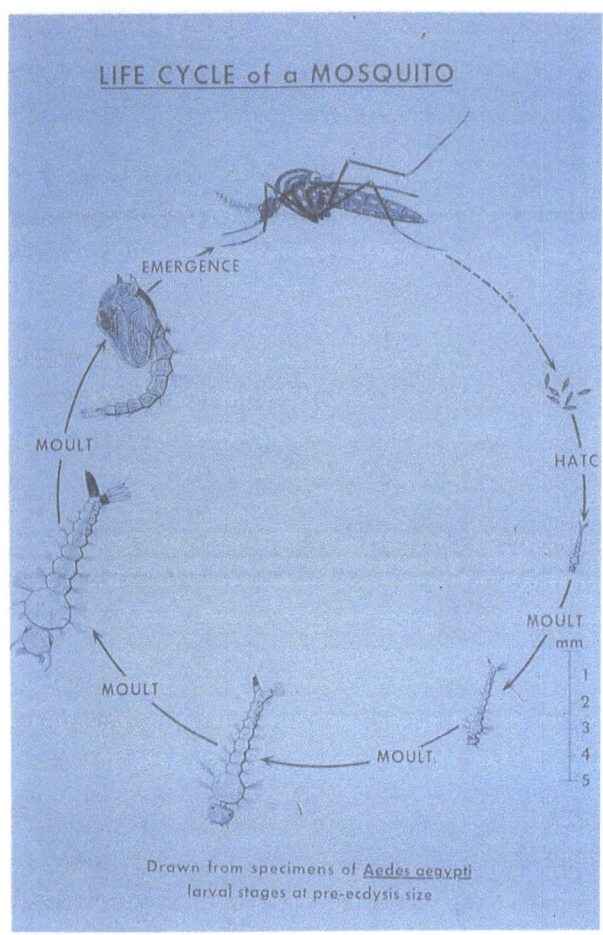

Fig. 2.6 **Life cycle of the mosquito.** The egg, larval and pupal stages are shown.

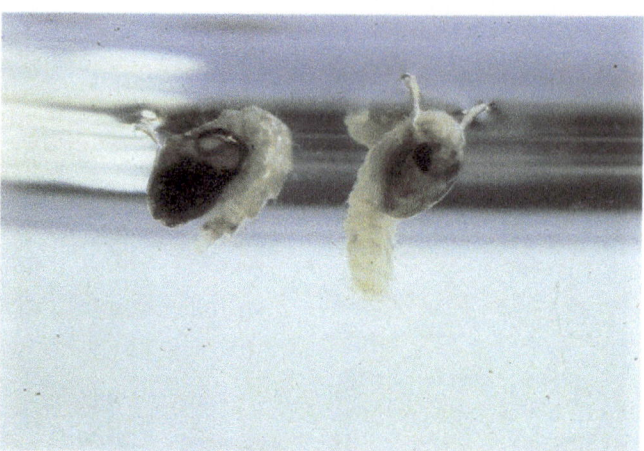

Fig. 2.8 Mosquito pupa.

Fig. 2.9 Female mosquito sucking blood.

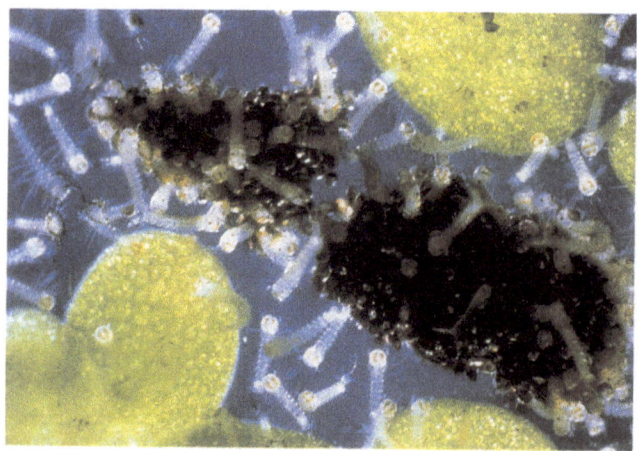

Fig. 2.7 Typical mosquito larva.

Fig. 2.10 *Anopheles* mosquito.

12

surface while the exoskeleton hardens and dries. Males will typically emerge first and swarm in the air over the breeding site. When the females emerge, mating will take place. The female mosquito stores sperm from a single mating in the spermatheca and these will be used to fertilize eggs from alternate ovaries when required. Thus the female mosquito will only need to mate once during her lifetime but may lay up to ten or so batches of eggs.

The female mosquito will typically feed (Fig. 2.9) in subdued light, especially during the night, but some species will readily feed in daylight. Depending on the species, female mosquitoes may rest indoors or outside before or after feeding. Identification of the species is thus essential in order that correct control measures may be carried out.

There are some 250 species of *Anopheles* mosquito (Fig. 2.10), about 100 of these are able to transmit malaria and 25 are notorious vectors in many parts of the world. Anophelines are usually pale brown in colour without any prominent markings. The wings typically have patches of darker scales, particularly along the leading edge, but the abdomen is not segmentally banded.

The maxillary palps are a pair of organs lying alongside the proboscis of a mosquito, between it and the antenna on either side. In the male anopheline mosquito (with bushy antennae), the palps are long and have a swollen or clubbed ending (Fig. 2.11).

In the female anopheline the palps are as long as, or longer than, the proboscis and lie close to it but are simple (without a clubbed ending) (Fig. 2.12). The resting position of the live anopheline mosquito is characteristic, with the head, thorax and abdomen in a straight line at an angle of about 45° to the resting surface (Fig. 2.13).

There are over 2000 species of culicine mosquito in 28 genera, the most important of which are *Aedes*, *Culex* and *Mansonia*. Culicine mosquitoes are typically more strikingly patterned than anophelines, with black and white or dark and light brown markings on the thorax, and dark and light segmental banding on the abdomen (Fig. 2.14). The wings, except in some species of the genus *Culiseta*, do not have patches of dark scales.

The palps in the male culicine are long but never clubbed, although they may have an elbowed or feathered ending (Fig. 2.15). The palps in the female culicine are short, usually only about one quarter of the length of the proboscis, and are never clubbed (Fig. 2.16). The culicine mosquito characteristically

Fig. 2.11 Clubbed palps typical of the male anopheline (*Anopheles*) mosquito.

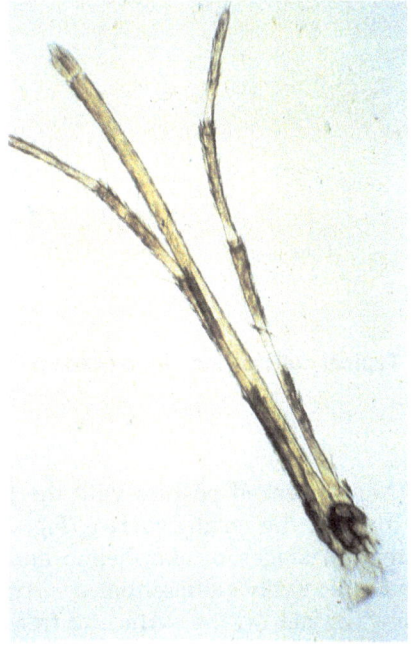

Fig. 2.12 Long simple palps of the female anopheline (*Anopheles*) mosquito.

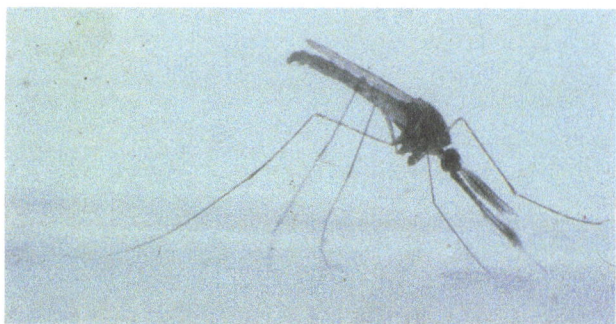

Fig. 2.13 Resting position of the anopheline mosquito (male).

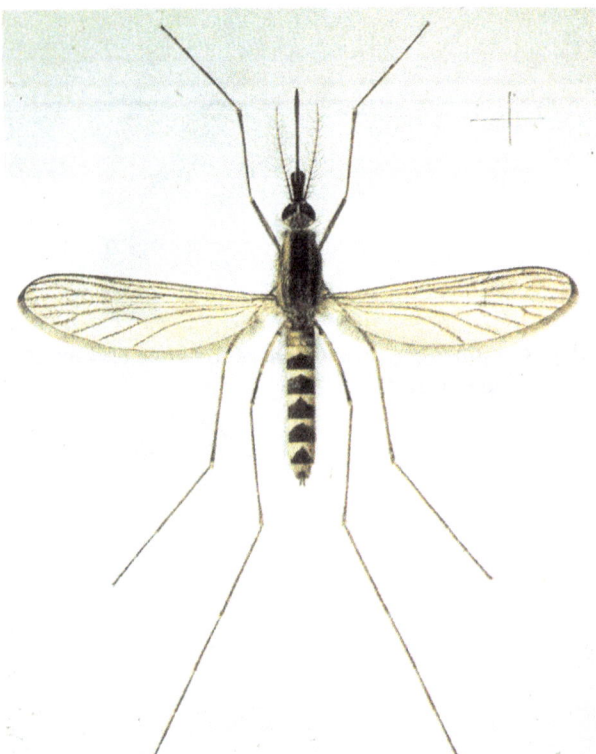

Fig. 2.14 Typical culicine mosquito (*Aedes*).

Fig. 2.15 Long palps of male culicine mosquito. These are never clubbed.

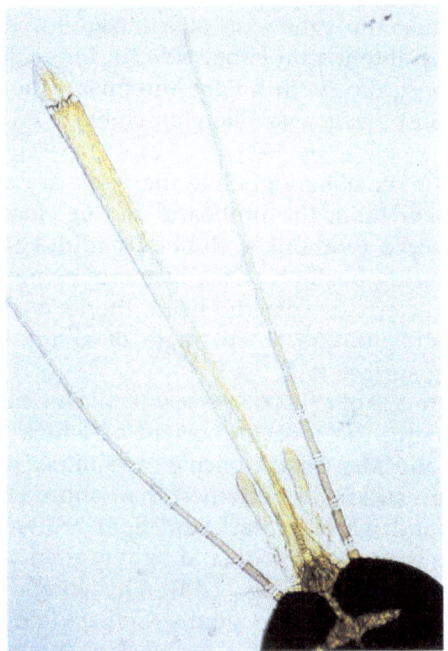

Fig. 2.16 Short palps of female culicine mosquito.

rests in a hunch backed posture with the head and abdomen towards the resting surface (Fig. 2.17).

The immature stages of anopheline and culicine mosquitoes are easily differentiated. Anopheline eggs are always laid on the surface of free water in batches of 20 to 30, often in a rosette pattern. Each egg is boat-shaped and has a pair of lateral air sacs (Fig. 2.18).

Culicine eggs never have lateral air sacs. The three genera that are of medical importance lay eggs in different conformations. *Aedes* females will lay eggs on a dry surface, which will eventually become covered in water (Fig. 2.19). This may take a wide variety of forms, for instance a stream bed that will become flooded in melt water in tundra regions. Eggs are often laid on the inside of water containers.

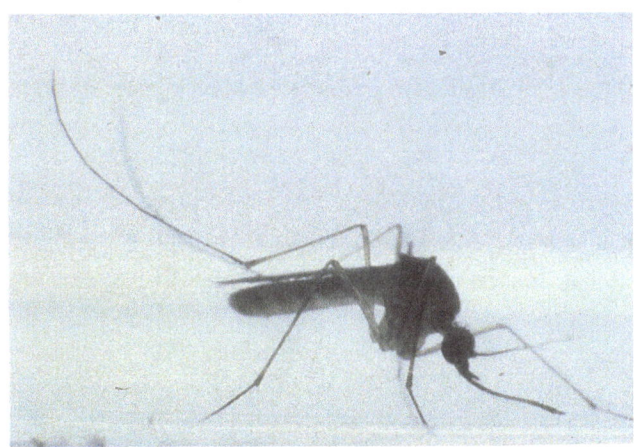

Fig. 2.17   Resting position of culicine mosquito. This is typically hunchbacked.

Fig. 2.20   *Culex* egg raft.

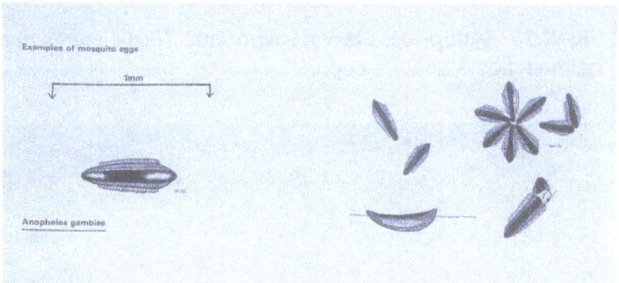

Fig. 2.18   *Anopheles* eggs.

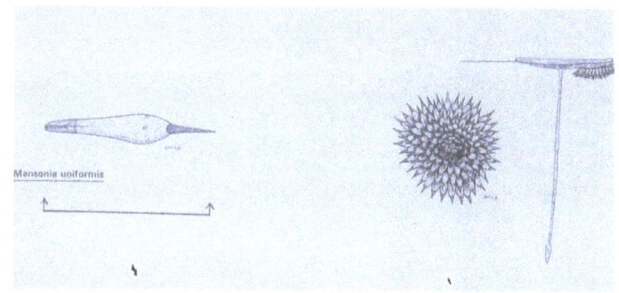

Fig. 2.21   *Mansonia* eggs.

Fig. 2.19   *Aedes* eggs.

The container will fill with rain water and cover the eggs, which will hatch within 1 to 6 hours. Only a proportion of the eggs hatch on the first wetting. *Aedes* eggs are able to resist desiccation for many months.

Culex eggs (Fig. 2.20) are laid on the surface of free water in a raft formation with 100 or more eggs stuck together side-by-side. The larvae will hatch through

a corolla or lid on the end that points down into the water. *Mansonia* eggs (Fig. 2.21) are equipped with a spike on one end, which is used to anchor the egg to vegetation in the water.

Anopheline and culicine larvae can be easily differentiated. The anopheline larva lies parallel to the water surface (Fig. 2.22a) in order to obtain air through the paired spiracles, which are flush with the ninth dorsal plate. The larva often feeds in this position, brushing small particles and microorganisms into the mouth with paired mouth-brushes on the head (Fig. 2.22b), which is typically narrower than the thorax (Fig. 2.22c). The head is rotated through 180° when feeding.

To assist adherence to the undersurface of the water by surface tension, the anopheline larva is equipped with paired palmate float-hairs and a waxy tergal plate on abdominal segments one to eight (Fig. 2.23).

The paired spiracles on the ninth segment of all culicine larvae are situated on the end of a dorsal extension or siphon (Fig. 2.24a). The larva will obtain air by fixing the tip of the siphon to the water

(a)

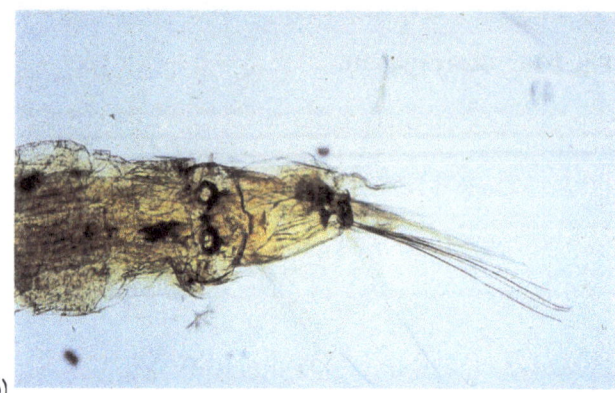

(b)

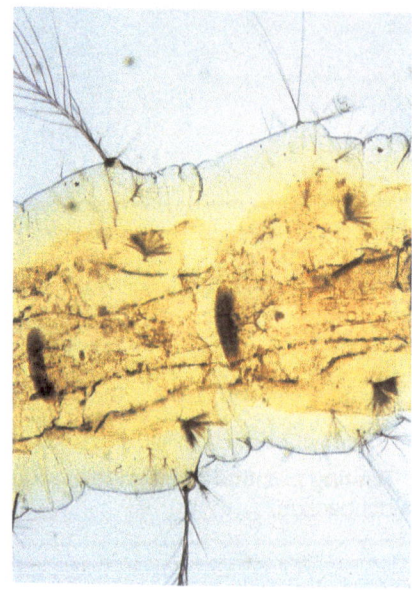

Fig. 2.23 *Anopheles* **larval segments.** Tergal plates and palmate float hairs are seen.

(c)

Fig. 2.22 *Anopheles* **larva.** (a) Lying parallel to the water surface. (b) Showing spiracular plate. (c) Showing paired mouth brushes.

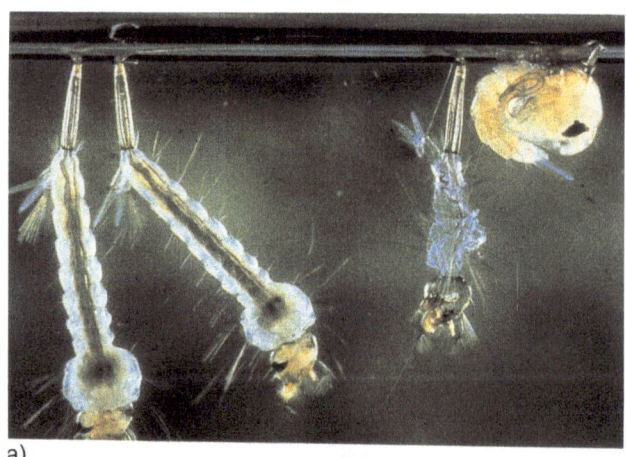

a)

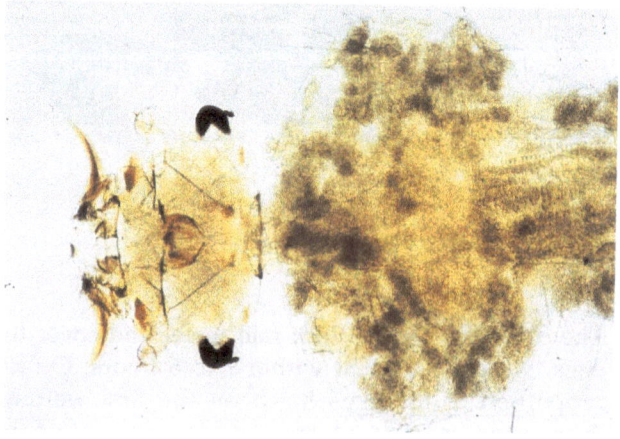

(b)

Fig. 2.24 **Culicine larva.** (a) Showing the tips of siphons at the water surface. (b) Head of larva, which is broader than long.

surface by surface tension and hanging vertically. Feeding in this position is difficult and is usually carried out below the surface, typically at the bottom. Thus culicine larvae do not have palmate float hairs, feathered bristles or waxy tergal plates (compared with anophelines). The head is broader than the thorax (Fig. 2.24b).

Fig. 2.25 *Aedes* **larval siphon.**

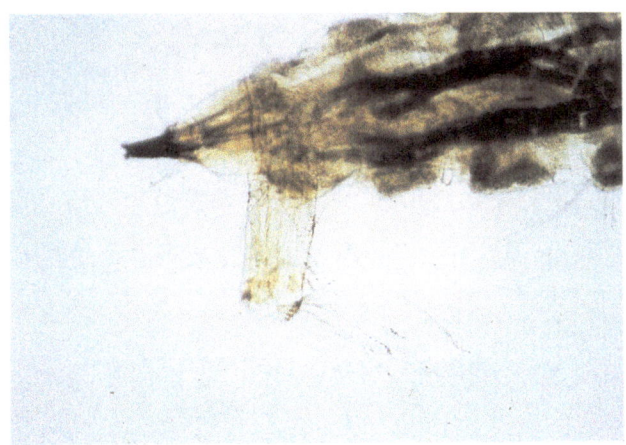

Fig. 2.27 *Mansonia* **larval siphon.**

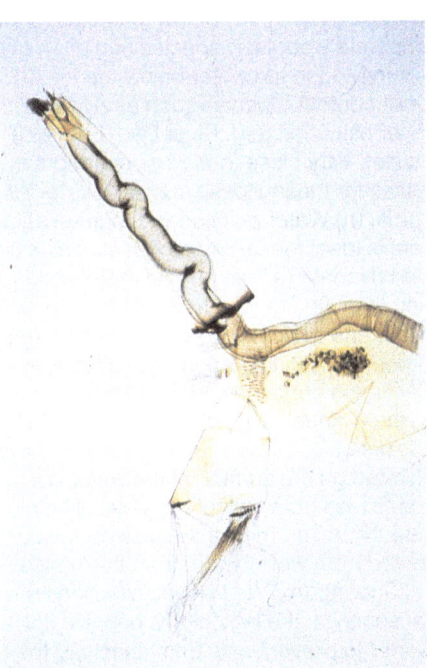

Fig. 2.26 *Culex* **larval siphon and tracheae.**

Fig. 2.28 **Larval siphon of** *Eretmopodites* **species.**

In the *Aedes* larva the siphon is short (Fig. 2.25), and with the tenth segment and the anal papillae, forms a Y-shaped ending to the abdomen. In the *Culex* larva the siphon is longer. In Fig. 2.26, the tracheae can be seen running from the spiracles down the siphon to the abdomen. In the *Mansonia* larva, a hook is present at the end of the siphon (Fig. 2.27) with which the larva fixes the spiracles to the stem of a water plant. Oxygen is thus obtained in this manner. The siphon in some other genera of mosquitoes may be excessively long (Fig. 2.28).

The larva is the feeding and growing stage of the mosquito. When the fourth stage or instar is fully developed the exoskeleton will be cast off to form the comma-shaped pupa. Pupae are very similar in all genera of mosquitoes, with minor differences only in the shape of the trumpets and the abdominal bristles (Fig. 2.29).

The immature stages of mosquitoes are always associated with free water of some sort but the type and location will vary considerably. Still water is preferred although slow-moving water at the edges

Fig. 2.29 Mosquito pupa.

of streams and rivers may be suitable. The larvae and pupae must be able to obtain water through the water surface, thus stagnant, encrusted water is not used. However, the water may be polluted with sewage, or even brackish. Figure 2.30 shows some typical and specific sites in which mosquito eggs, larvae and pupae may be found.

(a)

(b)

Fig. 2.30 Varied environments in which different species of mosquito are able to breed. (a) Free water with vegetation. *Anopheles gambiae* will breed in rice fields such as this in the Sudan. (b) Edges of large rivers (Belize). (c) Non-flowing areas of streams. Larvae and pupae will remain in these regions, not in the flowing areas. (d) Edges of flooded fields (Nepal). (e) Below the frozen surface of pools. Temperate woodland species can often overwinter as larvae, surviving in the water below the ice. (f) Water with high salt content. Species such as *Aedes detritus* breed in river estuaries (e.g. River Dee, England). (g) Brackish water, very close to the sea shore provides breeding sites for mosquitoes such as this *Aedes detritus* site in Cyprus. (h) Water polluted with human effluent. This environment is ideal for *Culex quinquefasciatus* as in this village pond (Nepal). (i) This puddle in Botswana supports two species of mosquito, one that prefers to develop in shaded water and the other in bright sunlight. (j) Humans provide breeding sites for various types for mosquito: rain water collecting in containers in a rubbish tip is a typical example, where *Aedes aegypti* may breed. (k) Drinking water containers also provide an ideal site. Polystyrene granules floated on the surface of the water will prevent female mosquitoes from laying eggs but will not contaminate the water. (l) Five species of mosquito larvae were collected from waste water from this house (Khartoum, Sudan). (m) Very small collections of water can support mosquitoes: the legs of this beehive in Nepal stand in tins of water to prevent ants from attacking the honey. Mosquito larvae were found in each tin. (n) The subfamily Megharininae pass through the immature stages in water collected in the axils of plants. The larvae are usually cannibalistic, thus only one larva is found in each axil. (o) In some species of *Aedes*, eggs will be laid by the gravid female on dry surfaces, which will become covered in water when rain falls or when snow and ice melt. This dry river bed in Cyprus floods in the winter, and eggs laid on dry soil and rock will hatch within a few hours of immersion. (p) *Aedes aegypti* were found in these fire buckets in Hong Kong. A weekly 'dry day', when all water containers are emptied, will break the breeding cycle. (q) *Anopheles* larvae were found in this marshy ground in southern Cyprus. (**Continued over**)

(c)

(d)

(e)

(f)

(g)

(h)

(i)

(j) **Fig. 2.30 continued.**

(k)

(l)

(m)

(n)

(q)

(o)

(p)

MEDICAL SIGNIFICANCE

The female mosquito typically requires a blood meal before each ovulation. The presence of a potential host is detected using several stimuli received by sensory organs, particularly in the antennae. These include changes in temperature and movement of air, and an increase in carbon dioxide, which is being expired by the host. The mosquito will land on the host and probe the tissue for a blood capillary. Only the stylet-like internal mouthparts are inserted, the outer sheath of the proboscis, the labium, being split dorsally and looped back.

To prevent the action of the host's blood-clotting mechanism, a small amount of anticoagulant saliva is injected by the mosquito. If the host is not equipped with the antibodies to react to this invasion of a foreign substance, all that will appear is a small reddish punctum mark (Fig. 2.31). After further bites, antibodies will be created and a reaction will occur at the site of the puncture. This may be immediate or delayed depending on the state of sensitization and will appear as a red and irritating swelling. Sensitization may occur within a few hours of the initial bite, but in some individuals it may take days or even weeks to occur. The antibodies are often specific for one species or a species group of

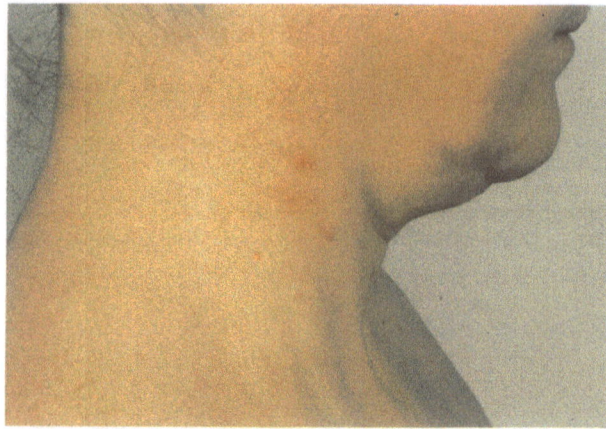

Fig. 2.31 Punctum resulting from mosquito bite in an unsensitized host.

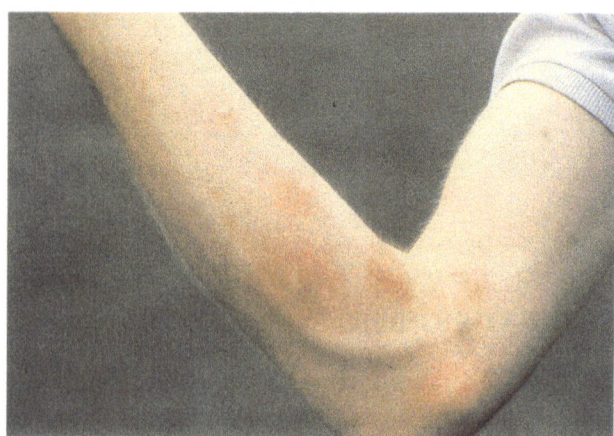

Fig. 2.32 Mosquito bites in sensitized host. This has resulted in swelling and redness with immediate or delayed irritation.

Fig. 2.33 Secondary infection of mosquito bites. (a) In the lower legs. (b and c) Gross infection of bites on the leg.

insects, and the rate of reaction can be very individual (Fig. 2.32).

After repeated attacks from the same species of insect, a state of immunity may be reached, when the host ceases to react to bites. Thus the state of sensitivity or immunity may explain, to a large extent, why some people appear to be bitten by insects more readily than others; it is more likely that all have been bitten but only some have reacted.

Irritating bites that are scratched and traumatized may often become infected with bacteria, particularly in a hot and humid climate. Secondary infection is frequently a problem when it occurs on the lower limbs (Fig. 2.33).

Mosquitoes can transmit a wide range of disease-causing organisms. These organisms can only develop through a particular phase of their life cycle if they are present in the mosquito at the right stage of the cycle. Many are specific to particular groups or even species of mosquito, which will thus act as an essential 'developmental' vector (Table 2.1).

Malaria

Malaria (Fig. 2.34) is transmitted in this way by several *Anopheles* species (Figs 2.10, 2.35 and Table 2.2). The gametocytes of the *Plasmodium* parasite are taken up in a blood meal from an infected human host and pass to the mid-gut of the female mosquito where they will mate; oocysts are formed on the outside of the mid-gut wall, developing inwards into the body cavity (Fig. 2.36).

When mature, the oocysts burst and sporozoites move through the body cavity (Fig. 2.37a), many thousands being chemically attracted to the paired trilobed salivary glands (Fig. 2.37b), from where they pass to a new host when the mosquito feeds again. This development period may be from 10 to 15 days depending on the species of *Plasmodium*. A female mosquito is termed 'infected' if oocysts are

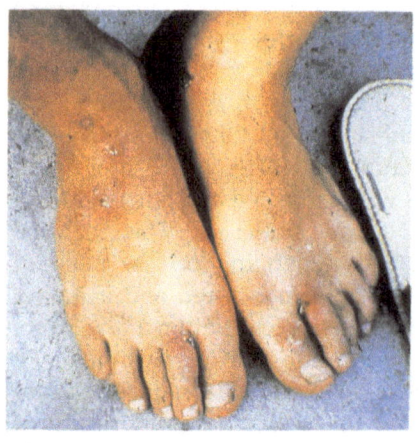

(a)

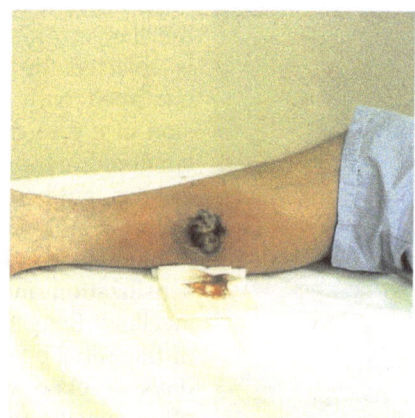

(b)

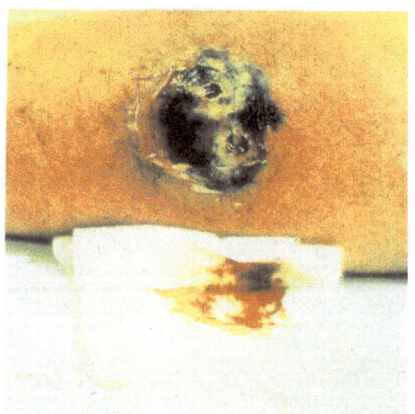

(c)

Table 2.1 Diseases transmitted by the two main groups of mosquitoes	
Anopheline	Culcine
Malaria	Filariasis *Wuchereria bancrofti* *Brugia malayi*

Filariasis
A few arboviruses
  e.g. O'nyong nyong
	Many arboviruses
	  e.g. Yellow fever
	    Dengue fever
	    Japanese
	      encephalitis

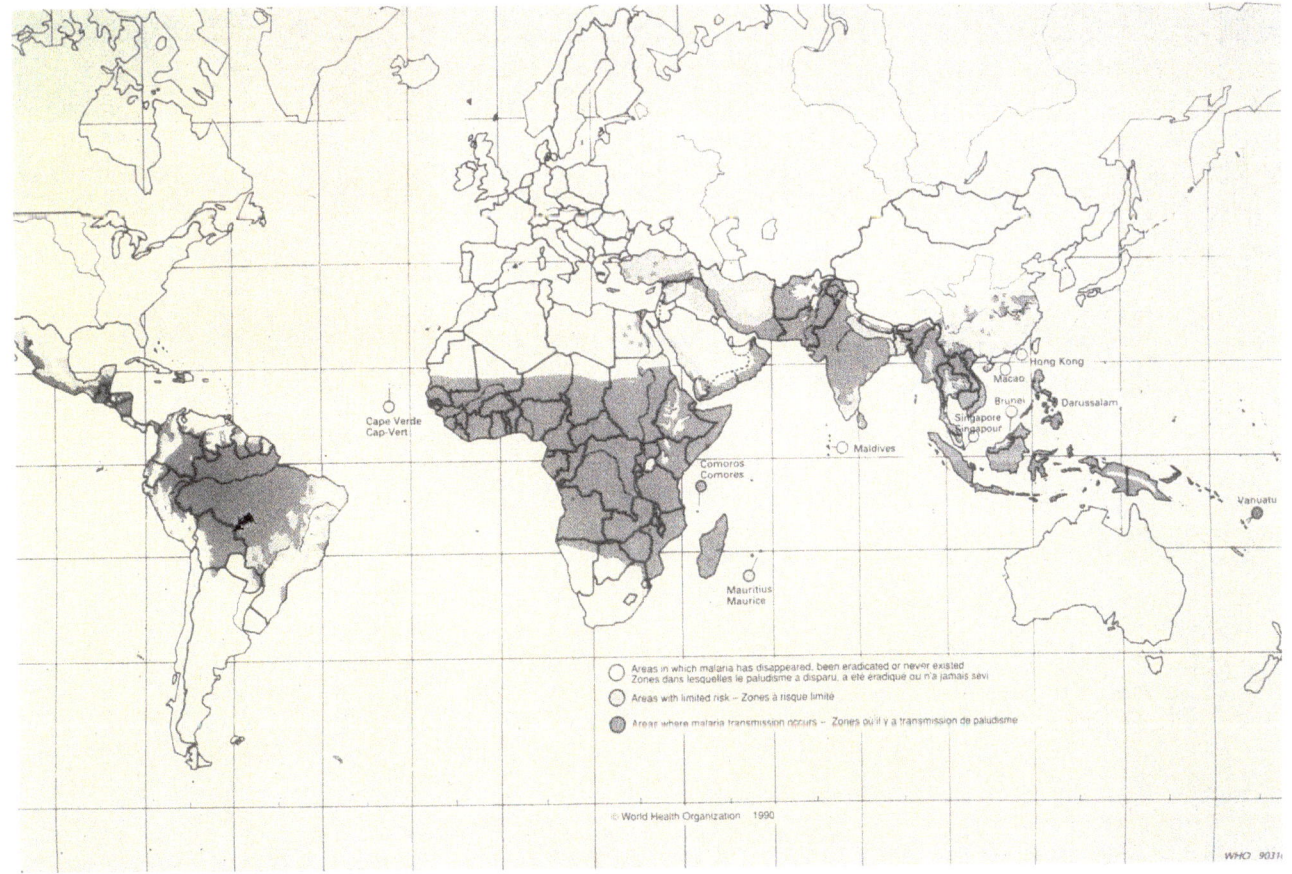

Fig. 2.34 Distribution of malaria.

Table 2.2   Examples of important malarial vectors

Anopheles albimanus *Anopheles darlingi*	Mexico, Central and South America
Anopheles gambiae *Anopheles funestus*	Africa, south of Sahara
Anopheles stephensi	Egypt to China
Anopheles sacharovi	Mediterranean and Middle East
Anopheles maculatus	India to China
Anopheles hyrcanus group	Mediterranean to Japan

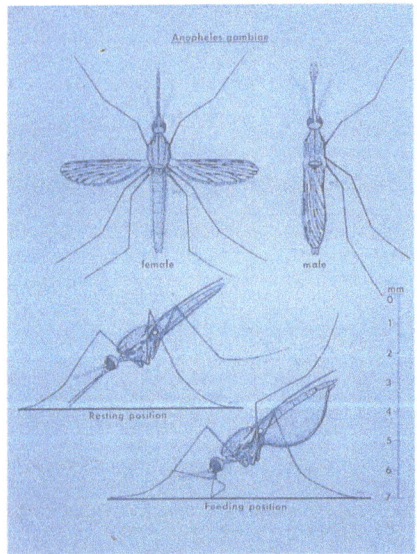

Fig. 2.35   *Anopheles gambiae*.

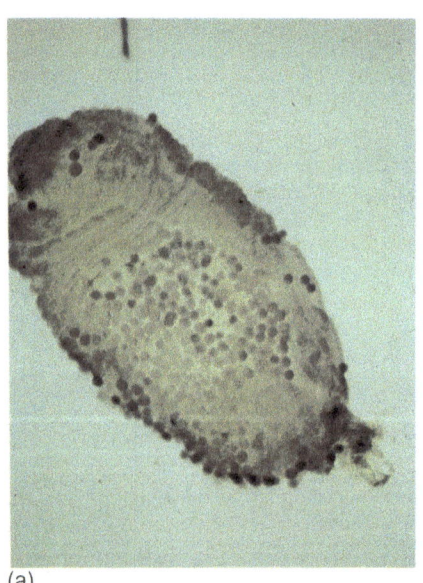

(a)

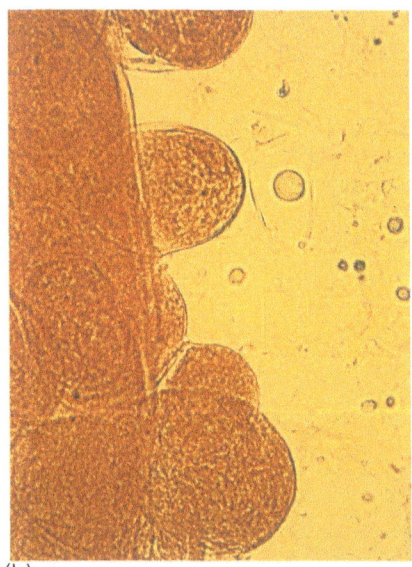

(b)

Fig. 2.36   Oocysts. (a) In the wall of the mosquito mid-gut. (b) Enlarged view.

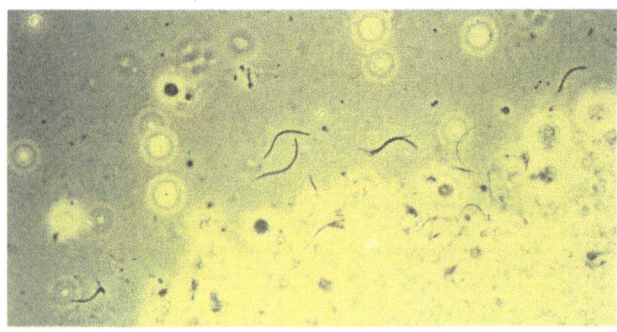

(a)

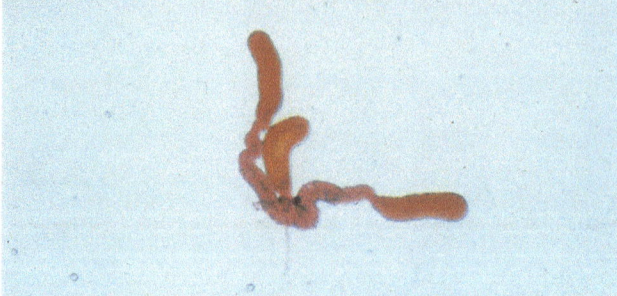

(b)

Fig. 2.37   Sporozoites. (a) Moving through the body cavity to infect (b) the salivary glands.

present in the mid-gut, but is only capable of transmitting malaria (i.e. it is 'infective') if sporozoites are present in the salivary glands. The 'infective rate' (percentage of infective mosquitoes caught) is normally below 1.0% in an area of low endemicity, rising to perhaps 5.0% in a highly endemic area.

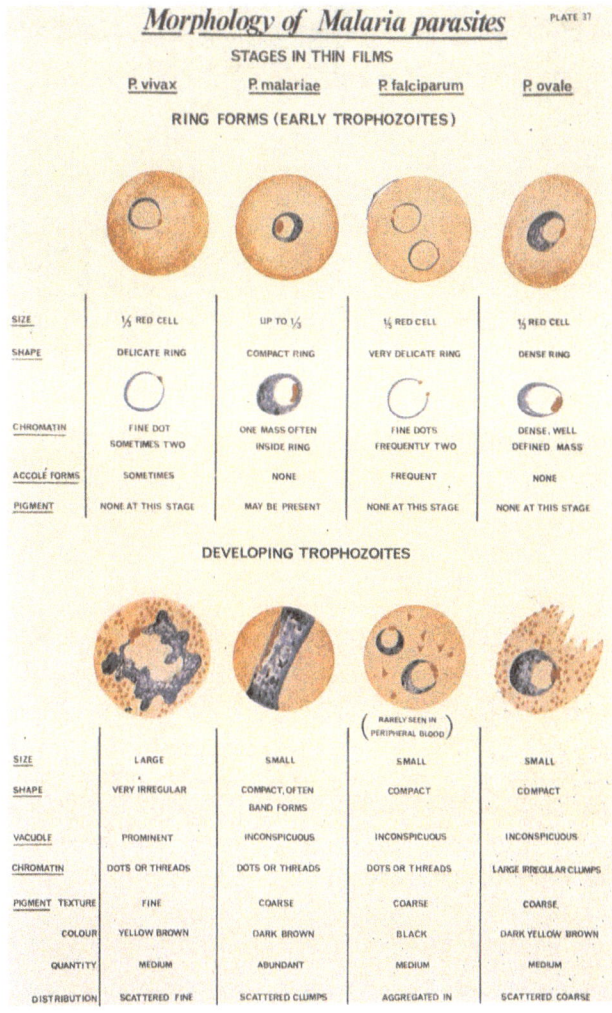

Fig. 2.38 Morphology of malaria parasites. (From Jeffreys and Leach.)

Clinical aspects

Four species of *Plasmodium* parasite (Fig. 2.38) cause illness in humans:

1. *Plasmodium falciparum* causes several million deaths annually in tropical populations, especially in children, and is a lethal infection in persons without immunity or protection. Parasitization of red blood cells leads to capillary sludging, which results in a marked reduction in oxygen delivery to all tissues, most dangerously in brain and kidney, and to fever and destruction of all infected red cells, causing progressive anaemia with jaundice and enlargement of the spleen. Death results most often from cerebral dysfunction, but severe shock, secondary bacterial septicaemia and hypoglycaemia may contribute.

2. *Plasmodium vivax* and *Plasmodium ovale* cause nonlethal recurrent fever, typically on every second day, with splenomegaly.

3. *Plasmodium malariae* also causes recurrent fever, but on every third day, and can also cause severe renal glomerular damage in young children, resulting in massive urinary protein loss and generalized oedema.

Malaria is diagnosed by demonstration of the parasites in stained blood films. Treatment is with chloroquine, except for the many strains of *P. falciparum* that are now partially or completely resistant to it, for which quinine remains the best treatment. Relapses of fever due to *P. vivax* and *P. ovale* malaria are prevented by the subsequent administration of primaquine.

Prevention of malaria depends on the control of mosquito populations, or measures to prevent mosquito bites, such as bed-nets, suitable clothing and chemical repellents, and on the administration of suitable chemoprophylactic drugs to those at risk.

Arboviruses

Wild animals will act as reservoirs for many of the mosquito-borne arboviruses, for example Japanese encephalitis (Fig. 2.39a). The virus may be found in water birds and is maintained by mosquitoes in the habitat.

Infected mosquitoes may feed on domestic animals (pigs in particular), which will act as efficient multiplication hosts and reservoirs. In parts of southeast Asia, where the disease is common, humans live in close contact with domestic animals, and domestic breeding sites are ubiquitous (Fig. 2.39b).

Infected mosquitoes will also readily feed on humans. Once the virus crosses from the vascular to the nervous system, mortality in untreated cases may be as high as 30%, with a further 30% suffering physical or mental impairment.

Clinical aspects

Arbovirus infection in humans may cause:

1. A mild short-term fever with muscle pains, e.g. dengue, sandfly fever, West Nile fever.
2. A brisk febrile illness with a maculopapular rash,

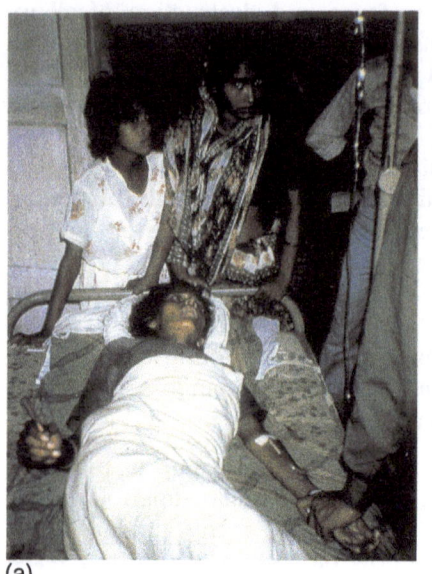
(a)

(b)

Fig. 2.39 Japanese encephalitis. (a) Patient suffering from the condition. (b) Pigs and chickens can act as reservoirs of Japanese encephalitis. Ample breeding sites are present (Nepal).

red eyes, joint pains and enlarged lymph nodes, e.g. Chikungunya, Ross River fever.
3. A severe illness with shock, a haemorrhagic rash and, in severe cases, jaundice (similar to yellow fever), e.g. dengue haemorrhagic fever (Fig. 2.40), Rift valley fever.
4. Encephalitis, e.g. Japanese encephalitis.

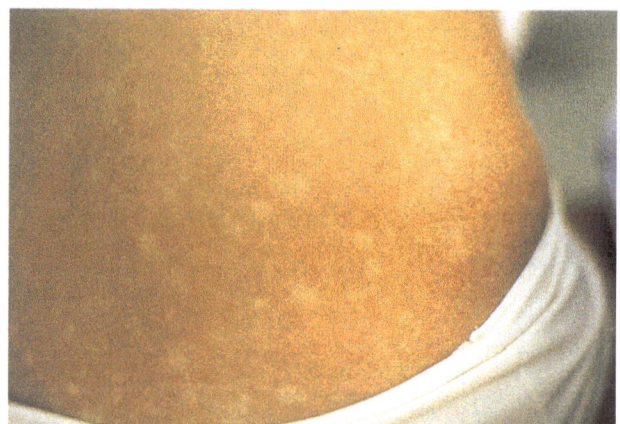

Fig. 2.40 Dengue rash.

Fig. 2.41 Distribution of yellow fever.

Diagnosis is made by serological methods. Treatment is generally symptomatic; haemorrhagic forms will require intravenous fluids. Some of these infections may respond to treatment with ribavirin. Production of vaccines for arbovirus diseases is generally difficult, although very effective vaccines exist to prevent Japanese encephalitis.

Yellow fever

Yellow fever (Fig. 2.41) is caused by an arbovirus transmitted by several culicine (*Aedes*) species, particularly the notorious urban vector *Aedes aegypti* (Fig. 2.42). The virus occurs in forest-dwelling monkeys in many parts of tropical Africa and Central and South America, where it is transmitted by several *Aedes* and *Haemagogus* mosquitoes.

Monkeys are not normally affected by the virus but if humans are bitten by these forest-dwelling mosquitoes the disease may be contracted. Thus sporadic cases of jungle or sylvatic yellow fever may occur in people living, travelling or working in endemic areas.

Urban epidemics of yellow fever occur when the infected host or infective vector find their way to a heavily populated area where *Aedes aegypti* is found (Fig. 2.43). These mosquitoes will breed in small collections of water such as that in water pots, tin cans and drainage gulleys (Fig. 2.44). This species is particularly anthrophilic (preferring human blood) and is a very efficient vector of yellow fever.

Mosquito control in urban yellow fever areas is often achieved by enforcement of 'dry days', when all water containers are emptied once a week thus breaking the mosquito breeding cycle. A fine is

(a)

(b)

Fig. 2.43 Urban, heavily populated areas encourage epidemics of yellow fever. (a) Typical populated area. (b) *Aedes aegypti* were breeding in this drain in Belize City, Central America.

Fig. 2.42 *Aedes aegypti*, the yellow fever or tiger mosquito. This species is easily identified by the silver lyre marking on the thorax.

Fig. 2.44 Typical *Aedes aegypti* breeding site. Note larvae present in the gulley.

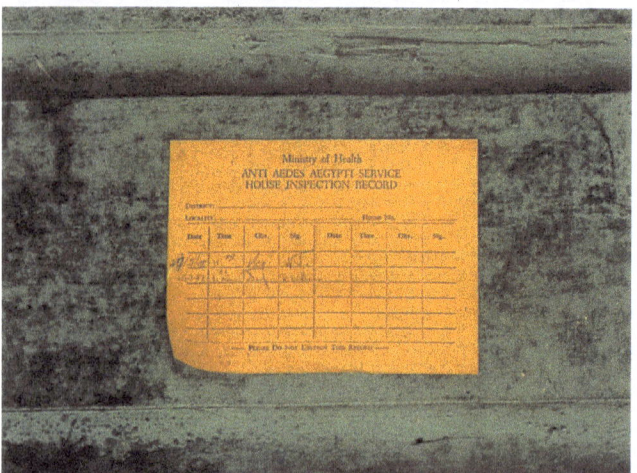

Fig. 2.45 Yellow mosquito inspection card.

imposed by health inspectors on the owner of a property on which mosquitoes are found breeding. A yellow card (Fig. 2.45) fixed to the building shows that it was free from potential vector mosquitoes on inspection.

Clinical aspects

Yellow fever usually occurs in epidemics. Many patients suffer only a short feverish illness for 3 to 4 days with headache and muscle pains. A minority will have a brief respite, then become seriously ill with a high fever, vomiting, severe headache and finally death from gastrointestinal haemorrhage or liver or kidney failure. Diagnosis is made serologically. Treatment is supportive, with intravenous fluids, blood transfusion, and steroids. Prevention is readily achieved by the use of the excellent vaccine, 17D.

Filariasis

Mosquitoes will also act as vectors of filarial worms, which are pathogenic to humans (Table 2.4). *Wuchereria bancrofti* and *Brugia malayi* have a nocturnal periodicity, being found in the peripheral blood of the host in the infective microfilarial form at night, whereupon they are taken up by mosquitoes feeding nocturnally. The worm pierces the gut lining and moves to the thoracic muscles where it develops and moults to the infective larval stage. It then moves to the hollow labium, piercing its way through the tip into a new host via the bite puncture when the mosquito feeds. This development takes 10 to 12 days. An infected mosquito rarely contains more than four or five developing larvae, since an overburden would impair the ability of the insect to fly and hence to further the life of the parasite. In the human host, the worm may obstruct lymphatic vessels causing gross enlargement of limbs (Fig. 2.46) and other organs such as the scrotum.

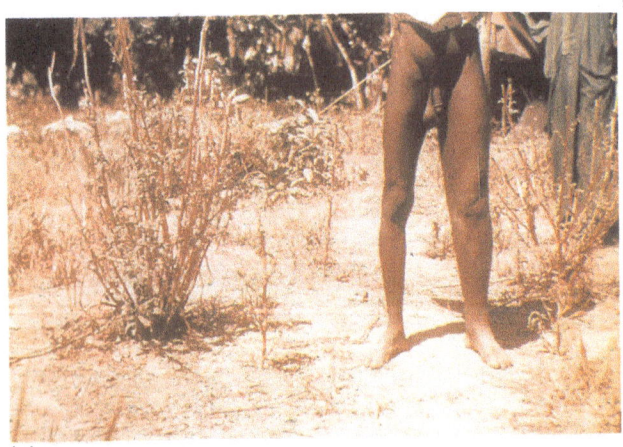

(a)

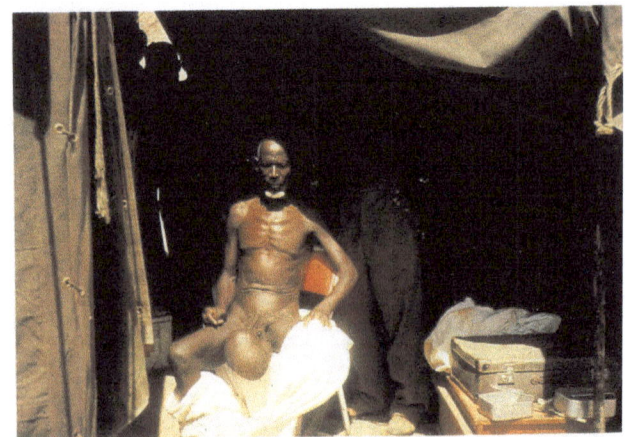

(b)

Fig. 2.46 Filariasis cases. (a) The leg. (b) The scrotum.

Table 2.3 Mosquito arbovirus vectors

Yellow fever		
(urban)	*Aedes aegypti*	Central and South
(sylvatic)	*Aedes simpsoni*	America, Africa
	Aedes africanus	south of Sahara
	Haemagogus spp.	
Dengue fever	*Aedes aegypti* and other *Aedes* spp.	Tropics
Encephalitis		
Japanese	*Culex fatigans*	Indian
	Culex gelidus	subcontinent, Southeast Asia
Murray valley	*Aedes* and *Culex*	South USA to
West equine	spp.	northern South
East equine		America
St Louis		
Venezuelan		
Chikungunya	*Aedes aegypti*	East and South Africa, Southeast Asia
O'nyong nyong	*Anopheles gambiae*	East Africa

Table 2.4 Important mosquito vectors of filarial worms

Wuchereria bancrofti	
Culex fatigans	Caribbean, Gulf to Pacific
(=*Culex quinquefasciatus*)	
Anopheles gambiae	Africa, south of Sahara
Anopheles sinensis	Southeast Asia
Brugia malayi	Indian subcontinent
Mansonia uniformis	Malaysia
Mansonia longipalpa	Southeast Asia
Anopheles sinenis	

Clinical aspects

Lymphatic filariasis (*W. bancrofti, B. malayi*): the disease processes caused by filarial worms result from the host's immune response to their presence in the lymphatic channels, which cause lymphatic blockage and chronic oedema of the affected part (elephantiasis). The clinical presentation may be with recurrent fever and inflamed lymphatic vessels and glands, orchitis followed by hydrocoele, or abscesses in limb lymphatics. Involvement of abdominal lymphatics may result in the passage of milky white urine (chyluria) or chylous ascites. Worms in the region of the eye may cause iritis, keratitis or glaucoma. In some racial groups (e.g. Tamil), an abnormal immune response in the lungs leads to cough and shortness of breath with eosinophilic lung infiltrates and peripheral blood eosinophilia (tropical pulmonary eosinophilia).

Diagnosis is made by the discovery of microfilariae in stained blood films or by serological methods. Standard treatment has been with diethylcarbamazine but this may be replaced by Ivermectin.

A summary of diagnostic differentiation between all stages of anopheline and culicine mosquitoes is shown in Fig. 2.47.

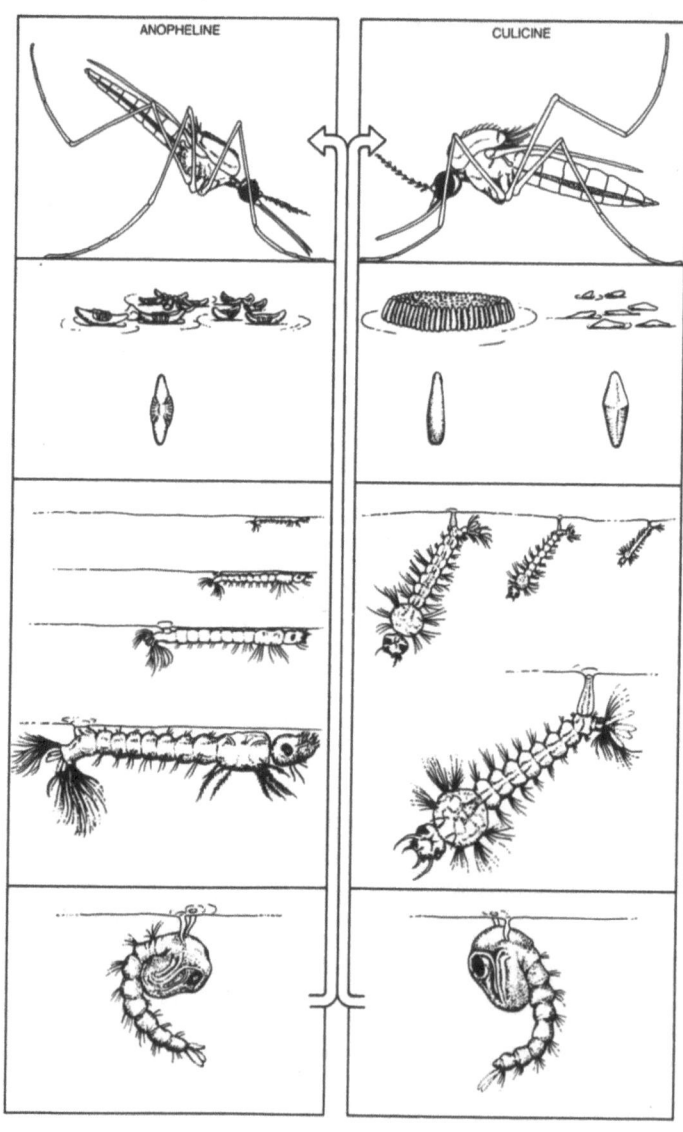

Fig. 2.47 Differentiation of anopheline and culicine mosquitoes.

3. Sandflies

INTRODUCTION AND DESCRIPTION

Sandflies are nematocerous Diptera and make up the subfamily Phlebotominae of the family Psychodidae. Also within this family are the non-biting owl-midges or moth flies in the genus *Psychoda*. Sandflies occur throughout the tropics, subtropics and warm temperate regions of the world, the genera *Phlebotomus* and *Sergentomyia* in the Old World and the genus *Lutzomyia* in the New World.

The sandfly is a small (2 to 4 mm in length), delicate midge-like fly (Fig. 3.1) with long thin legs and narrow pointed wings in which the second vein branches twice. At rest, the wings are held erect over the abdomen (Fig. 3.2). It is these features that differentiate sandflies from the non-biting Psychodids.

The proboscis of the sandfly is short and downwardly projecting, as are the longer pendulous palps. The antennae are long and similar in both sexes, although only the female bites and sucks blood, thus acting as a disease vector. The female abdomen is rounded at the end with small paired cerci (Fig. 3.3).

The male sandfly (Fig. 3.4) does not bite. The abdomen terminates in large paired claspers. The eyes in both male and female are dark and conspicuous, and the body and wings are covered in long hairs, rendering the sandfly an inefficient flier. Thus it will not be found far from its breeding site, nor will it fly in breezy conditions or much above ground level.

Fig. 3.1   Sandfly.

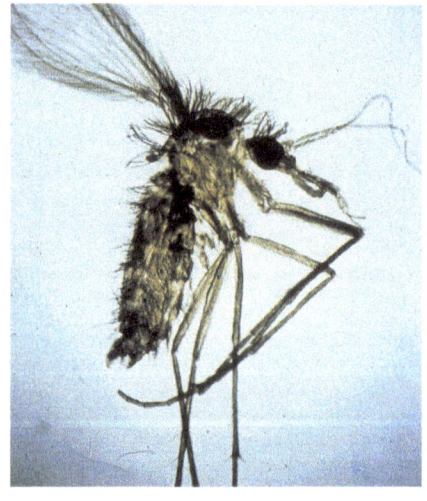

Fig. 3.3   Female sandfly.

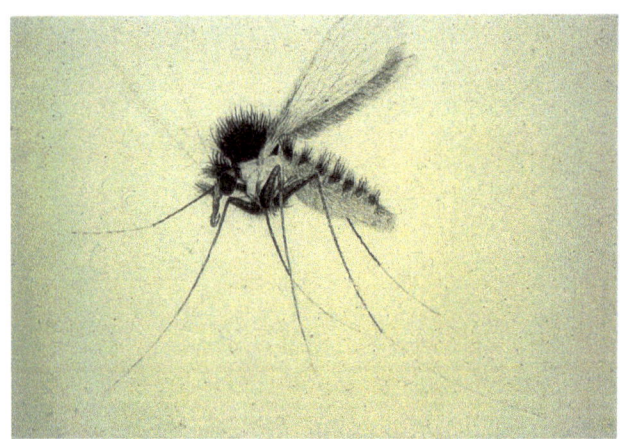

Fig. 3.2   Sandfly in the resting position.

Fig. 3.4   Male sandfly. In this slide preparation most of the body hairs have been removed.

LIFE CYCLE AND BREEDING SITES

There are some 600 species of sandfly that can be found in a wide range of habitats wherever the optimal breeding conditions of high temperature and humidity occur, often as a microclimate in leaf mould on the forest floor or in the cracks in bark and so on in tropical rain forests, where distribution of the sandfly may be particularly localized. Sandflies will also breed in arid climates, provided that a microclimate of high temperature and humidity is available (Fig. 3.5). Rock fissures and caves provide ideal sites for the immature stages. Female sandflies will normally only feed nocturnally or in subdued light.

Sandflies are small enough to pass through the mesh of a standard mosquito net. To prevent this, the net should be impregnated with permethrin (Fig. 3.6). The bite of the female sandfly is typically sharp and painful, and often causes considerable irritation.

The life cycle of the sandfly is one of complete metamorphosis (Fig. 3.7). The eggs (Fig. 3.8) are minute and are laid in cracks and crevices in the environment where the microclimate is high, the female laying several batches of ten to 100 eggs during her lifetime.

The minute maggot-like first-stage larva (Fig. 3.9) can be identified by the presence of one pair of long tail (caudal) bristles, which can be seen through the egg shell during development. The larva increases in size by feeding on microorganisms in its environment, including cast larvae and pupal skins. Second, third and fourth-stage larvae are progressively larger (Fig. 3.10) and have two pairs of caudal bristles (Fig.

(a)

(b)

Fig. 3.5 Breeding habitats of the sandfly. (a) Sandflies were breeding in the cracks and crevices in the mud walls of this building in Sudan, and feeding on the occupants. As the insect is a weak flier, the occupants avoid being bitten by traditionally sleeping outside when there is a breeze, or above ground floor level such as the roof. (b) During the day sandflies will shelter in a high microclimate such as that provided by the spaces in these stone walls in Greece.

Fig. 3.6 Mosquito net impregnated with permethrin.

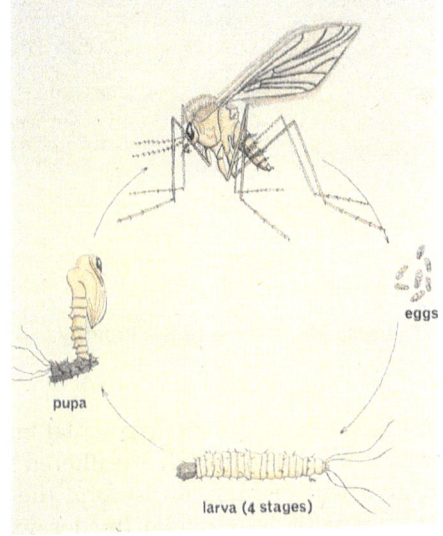

Fig. 3.7 Life cycle of the sandfly.

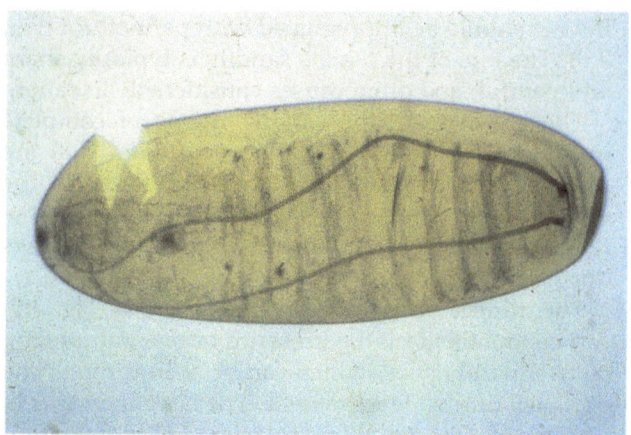

Fig. 3.8   Sandfly egg.

Fig. 3.9   First stage larva of the sandfly.

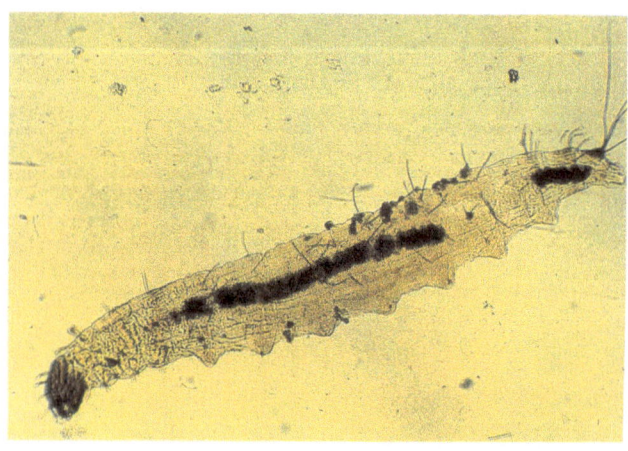

Fig. 3.10   Later stage larva of the sandfly.

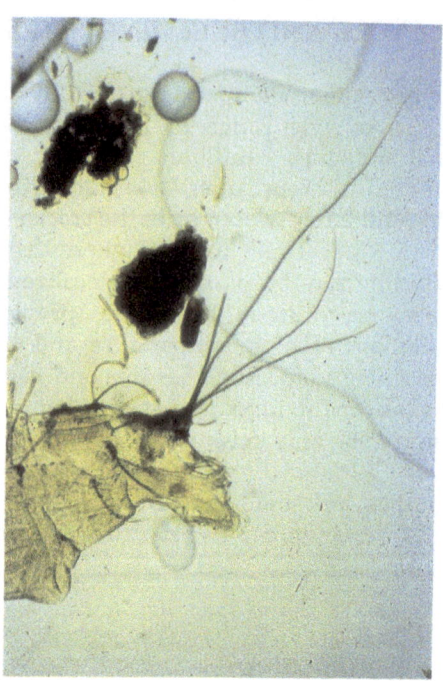

Fig. 3.11   Caudal bristles of mature sandfly larva

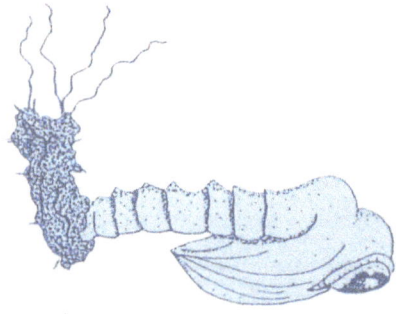

Fig. 3.12   Sandfly pupa.

MEDICAL SIGNIFICANCE

Female sandflies will act as vectors of three diseases: leishmaniasis, sandfly fever and bartonellosis.

Leishmaniasis

Leishmaniasis is named after Sir William Boog Leishman, Director General Army Medical Services 1923–1926.

Leishmaniasis is caused by species of the protozoan *Leishmania* and has a wide distribution (Fig.

3.11). All larvae have minute segmental 'matchstick' hairs, which are clubbed, with a feathered shaft. The mature larva will cast its skin to form the pupa but the cast skin with four caudal bristles will remain fixed to the end of the pupa (Fig. 3.12).

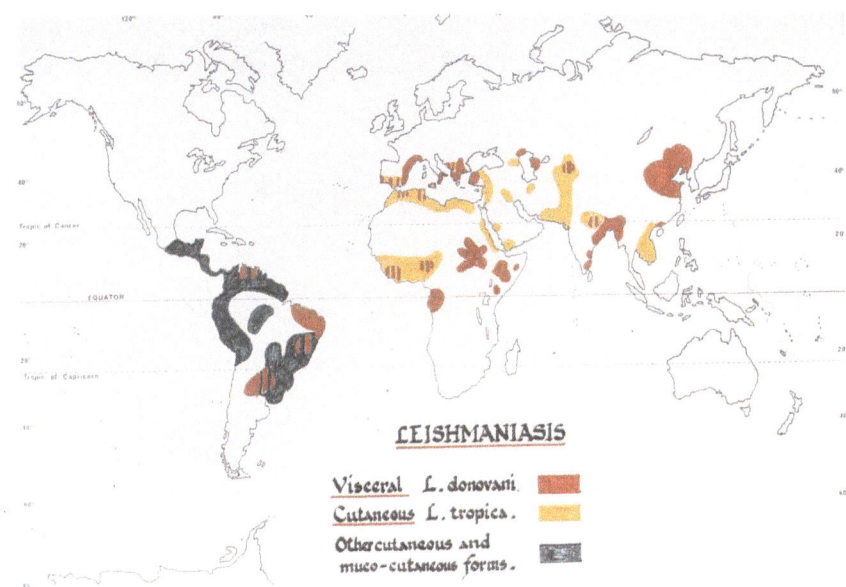

Fig. 3.13   Distribution of leishmaniasis.

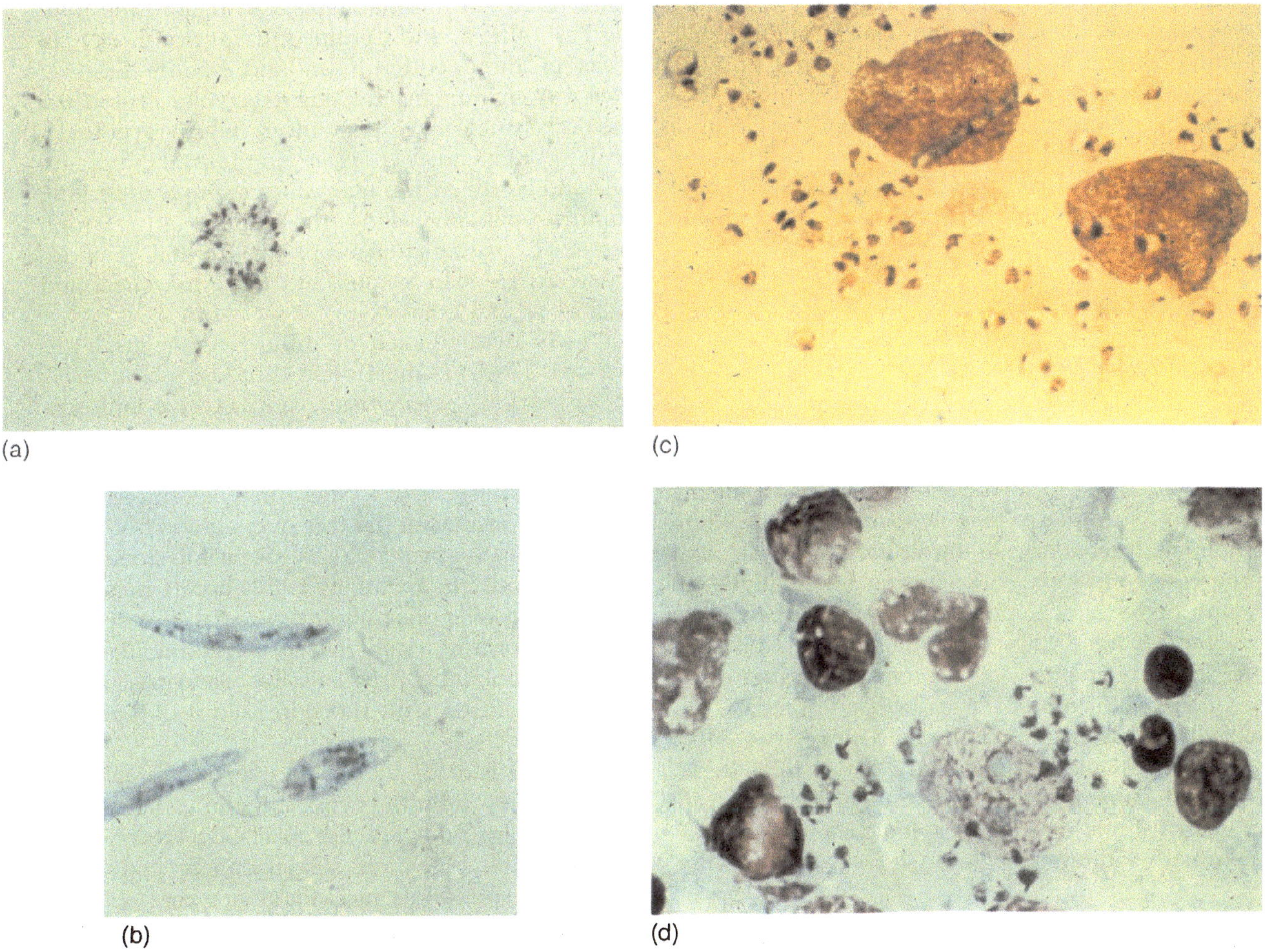

Fig. 3.14   Developing leishmaniasis. (a) and (b) Promastigotes; (c) and (d) Amastigotes ((c) in bone marrow; (d) in spleen).

Table 3.1 Important vectors of leishmaniasis	
Visceral	
Phlebotomus argentipes	Middle East
Phlebotomus chinensis	China
Phlebotomus perniciosus	Mediterranean
Cutaneous	
Phlebotomus sergenti	India
Phlebotomus papatasii	Mediterranean
Mucocutaneous	
Phlebotomus intermedius	Central and South America

3.13) and occurs in three distinct clinical forms: visceral (kala-azar), dermal (cutaneous) and mucocutaneous (espundia). Each is transmitted by particular species of sandfly in different regions (Table 3.1). *Leishmania* are taken up by the female sandfly while feeding from an infected host, which can be a wide range of animals including rodents and dogs. The organism undergoes development in the mid-gut of the sandfly over a period of 4 to 12 days, depending on the species, and is transmitted in a subsequent blood meal after migrating to the pharynx and proboscis (Fig. 3.14).

Clinical aspects

Visceral leishmaniasis (kala-azar)

Multiplication of *L. donovani* in the bone marrow leads to chronic fever, with progressively reduced output of all the formed elements of blood, causing infection (secondary to leucopaenia), a bleeding tendency (secondary to thrombocytopaenia), and progressive anaemia. The parasite also multiplies in lymph nodes, spleen and liver, which subsequently become enlarged. The patient becomes debilitated and hyperpigmented. Death usually ensues from secondary infection or haemorrhage. Patients infected with human immunodeficiency virus (HIV) suffer a particularly aggressive form of this illness. Diagnosis depends on demonstration of the parasite in stained bone marrow or splenic aspirates. Treatment with sodium stibogluconate or Paromomycin.

Cutaneous and mucocutaneous leishmaniasis

At least ten different species or subspecies of *Leish-* *mania* can cause lesions of the skin or at the junction of the skin with the nasal mucosa, and different reservoirs of infection exist. This spectrum of diseases occurs in the Mediterranean region, the Middle East, Africa, and Central and South America. In areas of the Mediterranean and Middle East, *L. tropica* from human and dog reservoirs causes the classical, single, dry skin ulcer, which eventually heals spontaneously. *L. major*, in the Arabian peninsula and North Africa, may cause multiple ulcerating papules, possibly along the course of a lymph vessel. *L. mexicana mexicana* in Central America causes single skin lesions, but these may heal and relapse several times.

Widely disseminated or diffuse cutaneous leishmaniasis (DCL) is most often caused by *L. aethiopia* or *L. mexicana amazonensis*, and erosive mucocutaneous ulcers, which are susceptible to secondary bacterial infection (espundia), are caused usually by *L. braziliensis braziliensis*. These may become chronic destructive lesions of the face over a period of many months. Finally, post kala-azar dermal leishmaniasis (PKDL) should be mentioned; this occurs mainly in India, often after treatment of visceral disease, and takes the form of a generalized maculonodular skin infiltrate containing transmissible parasites, which may be confused with the skin lesions of lepromatous leprosy.

Diagnosis is made by the discovery of parasites in material taken from the active edge of a skin lesion. Serological methods are also available. Treatment is with parenteral sodium stibogluconate and some parasites respond to topical Paromomycin.

Species of *Leishmania* causing dermal or cutaneous leishmaniasis will multiply around the site of the sandfly bite causing the typical sore, which may spread and disfigure the patient (Fig. 3.15). Certain species of *Leishmania* in the New World will attack the mucous membranes and cartilaginous areas as well as the skin. Exposed areas of the body exposed to bites such as the ears are common sites of attack (Fig. 3.16). In untreated cases, the parasite may spread and cause considerable disfigurement.

Sandfly fever

Sandfly fever occurs in the Mediterranean region, the Middle East, Pakistan and northern India (Fig. 3.17). It is an acute febrile illness of sudden onset, and symptoms include a red face, severe headache and painful neck muscles. The limbs feel stiff but there is no skin rash. Diagnosis is serological and treatment symptomatic. Full recovery occurs in all cases, but a few patients suffer a second shorter illness after an interval of 2 to 3 days ('saddleback fever'). Long-lasting immunity is conferred after the first attack. The virus may be passed from one generation of sandfly to the next transovarially (i.e. through the ovaries) or through the cast larval skins and dead adults that are eaten by larvae. In the Mediterranean the vector is *Phlebotomus papatasii*.

Bartonellosis

Bartonellosis occurs only in the western Andes (Fig. 3.18). *Bartonella bacilliformis* multiplies in the red blood cells causing a rapid haemolytic anaemia, and in bone marrow, liver, spleen and lymph nodes. This 'oroya fever' is accompanied by bone pain and anaemia; survivors may subsequently develop a generalized warty skin rash or nodules ('verruga peruana'). Diagnosis is made from stained blood films, which show the bacilli within the red cells. Treatment is with chloramphenicol, co-trimoxazole or tetracycline. The vector is *Phlebotomus verrucarum*.

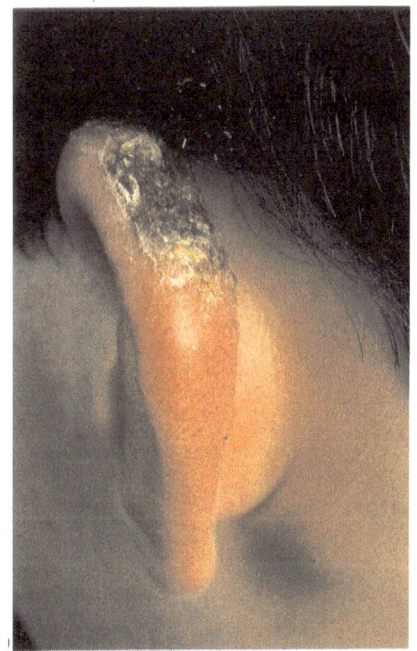
(a)

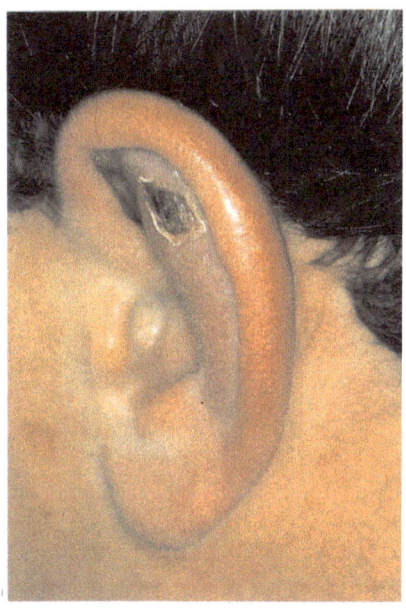
(b)

Fig. 3.16 Mucocutaneous leishmaniasis. (a and b) Lesions on the ears.

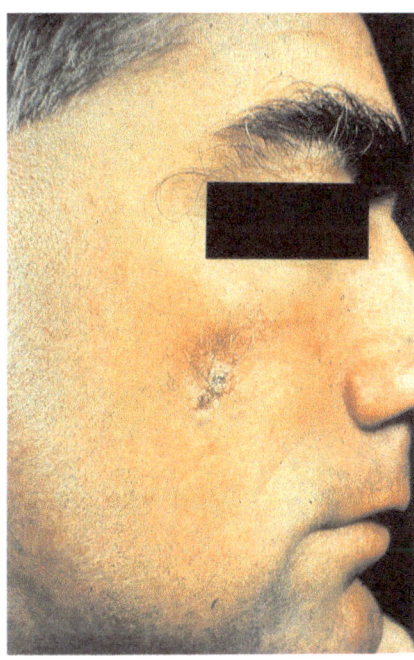

Fig. 3.15 Typical sore seen in cutaneous leishmaniasis.

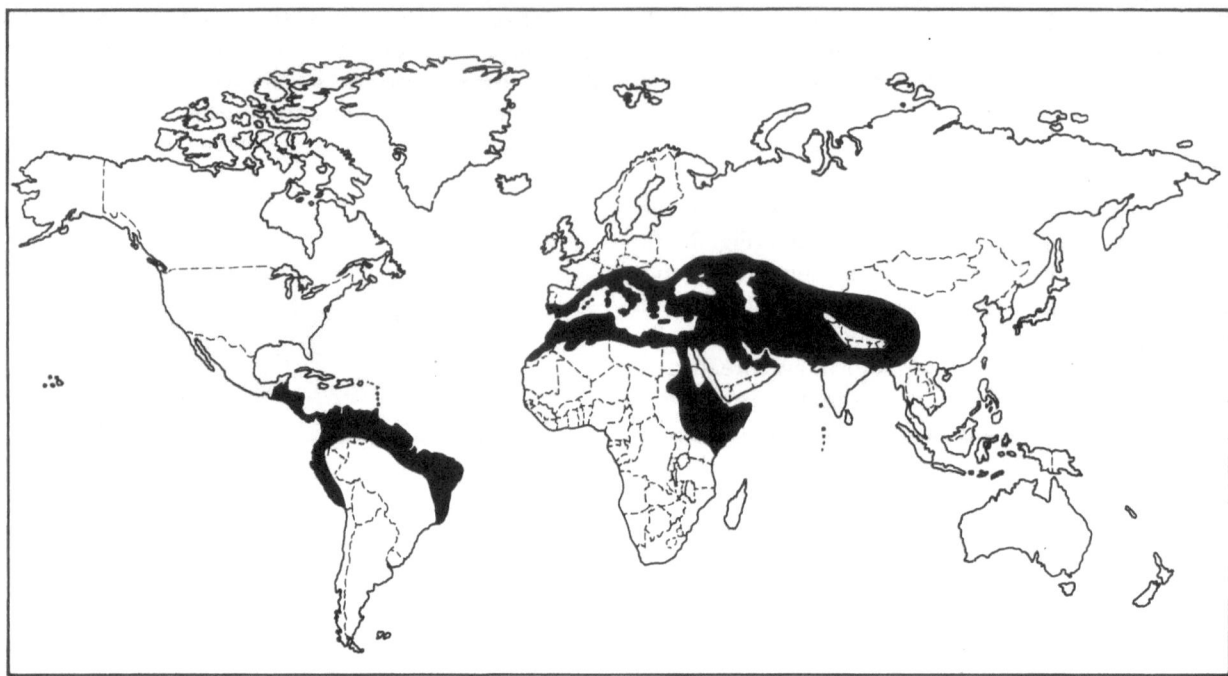

Fig. 3.17 Distribution of sandfly fever.

Fig. 3.18 Distribution of bartonellosis.

4. Biting midges
(Culicoides)

INTRODUCTION AND DESCRIPTION

Several genera within the nematocerous family Ceratopogonidae include blood-sucking midges (Fig. 4.1), sometimes misleadingly known as 'sandflies', particularly the genus *Culicoides* in which there are at least 800 decribed species. *Culicoides* range in size from 1 to 5 mm and, although small, they are fairly robustly built. They are usually dark in colour, have long bead-like antennae of 13 to 15 segments, a short downwardly-projecting proboscis and palps, and mottled wings with a characteristic mask-like costal cell on the leading edge.

Fig. 4.2 Female *Culicoides* after taking a blood meal.

Fig. 4.1 *Culicoides*.

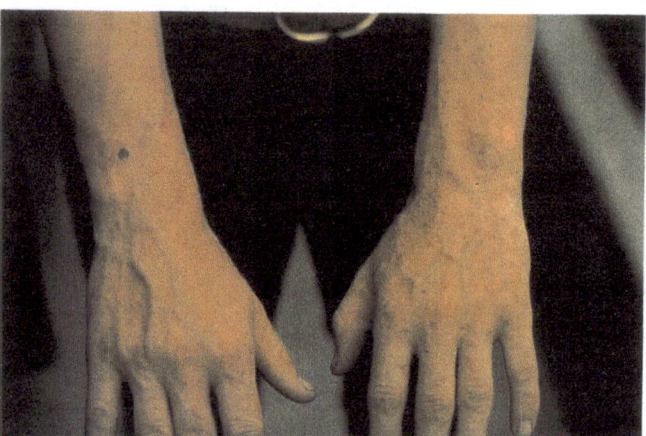

Fig. 4.3 *Culicoides* bites.

Central and South America and the Caribbean the filarial worm *Mansonella ozzardi* is transmitted. However, these worms are not thought to be pathogenic to humans, although they often occur with other pathogenic worms (Table 4.1).

LIFE CYCLE AND BREEDING SITES

Culicoides has an almost worldwide distribution wherever suitable conditions exist for the immature stages (Fig. 4.4). These stages require vegetable material with a high moisture content and breeding sites include temperate forests with rotting leaf mould, temperate peat bogs, tropical coastal swamp, beaches on which rotting seaweed has collected, and small tropical islands.

The life cycle is one of complete metamorphosis (Fig. 4.5). Eggs are laid in material with a high vegetable content, ranging from liquid mud to damp

Only the female *Culicoides* (Fig. 4.2), with sparsely haired antennae, feed on blood; males, with plumose antennae, only feed on vegetable fluids.

Female *Culicoides* will feed on exposed skin, particularly in sultry weather and may cause considerable irritation (Fig. 4.3), attacking, in large numbers, especially the scalp, face and hands. In parts of tropical West Africa, this midge may transmit the filarial worms *Mansonella* (=*Dipetalonema* or *Acanthocheilonema*) *perstans* and *Mansonella* (=*Dipetalonema*) *streptocerca*, while in parts of tropical

Table 4.1	Culicoides vectors of filarial worms	
Filarial worm	Vector	Region
Mansonella perstans	*Culicoides austeri*	West Africa
Mansonella streptocerca	*Culicoides grahami*	
Mansonella ozzardi	*Culicoides furens*	Caribbean

(a)

(c)

(b)

(d)

Fig. 4.4 Breeding habitats of *Culicoides*. (a) Temperate forest in Scotland. (b) Peat bog. (c) Seaweed on beach (Belize). (d) Tropical island (Caribbean).

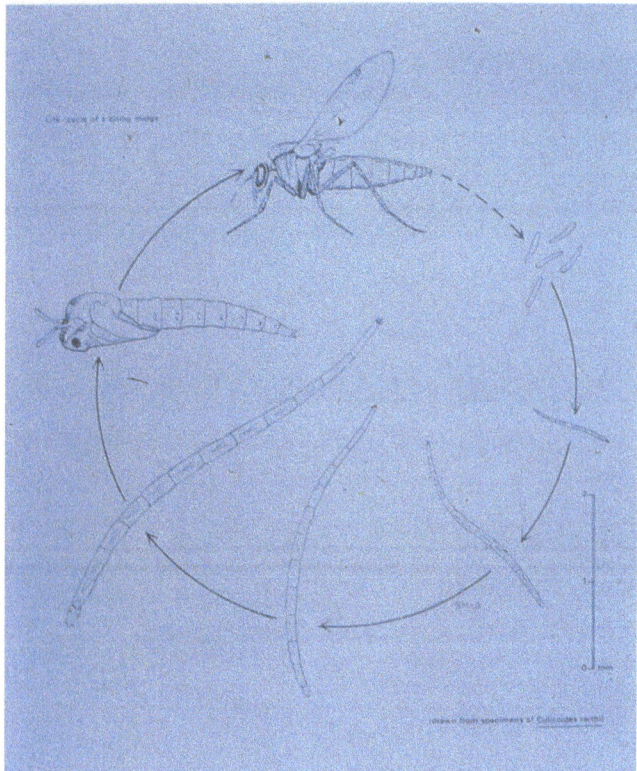

Fig. 4.5   Life cycle of *Culicoides*.

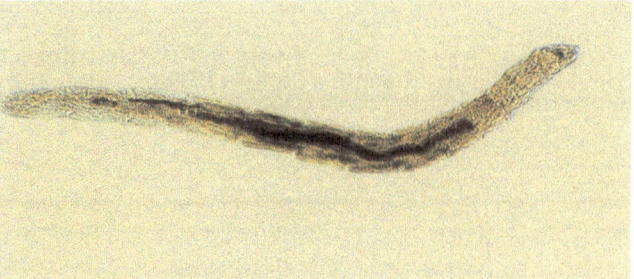

Fig. 4.6   *Culicoides* larva.

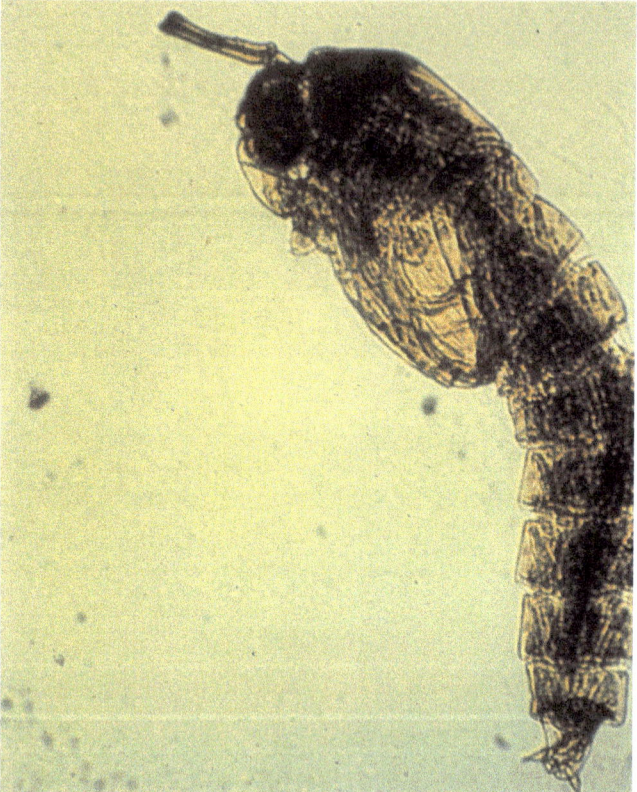

Fig. 4.7   *Culicoides* pupa.

rotting leaves, depending on the species. In the tropics the eggs hatch in 2 to 8 days, but may overwinter in colder regions. The larva is eel-like (Fig. 4.6) and feeds on decaying organic matter, surviving on oxygen dissolved in moisture, through the skin and via retractile anal gills. It moves through moist or wet media with a writhing motion.

The mature fourth-stage larva which is 6 to 7 mm long will cast its skin to form the comma-shaped pupa (Fig. 4.7), not unlike a mosquito pupa except that it is smaller. The pupa is slowly mobile but does not feed. After 3 to 7 days the adult emerges through a split across the dorsal surface.

MEDICAL SIGNIFICANCE

Culicoides midges transmit the filarial parasites *Mansonella ozzardi*, *Mansonella perstans*, and *Mansonella* *streptocerca*, which may be found in human blood in the form of microfiliariae. No specific pathology results from these infections, except that adults of *M. perstans* may occasionally cause retroperitoneal fibrosis and hence blockage of the ureter, leading to renal failure. *Culicoides* species may also be involved in the transmission of certain viruses, for example oropouche fever (Brazil), Rift valley fever (Africa, from the Rift valley across the Sahel) and eastern equine encephalitis (eastern USA).

5. Biting blackflies (Simulium)

INTRODUCTION AND DESCRIPTION

The family Simuliidae contains several genera, one of which, *Simulium*, is an important human-biting fly, known as the biting blackfly or buffalo fly (Fig. 5.1). *Simulium* is a small (2 to 5 mm), stoutly built hump-backed fly with short cigar-shaped pilose but well-segmented antennae (nine to 11 segments) in both sexes. The wings of *Simulium* are broad and

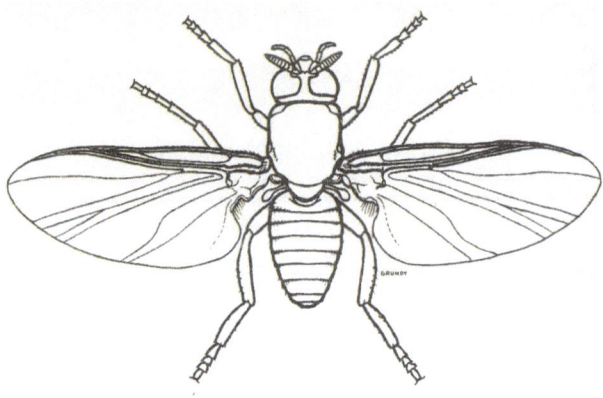

Fig. 5.1 *Simulium.*

Fig. 5.2 *Simulium* wing.

clear with the venation characteristically concentrated along the leading edge (Fig. 5.2). The compound eyes of the male are close together (holoptic), while those of the female are more widely spaced (dichoptic).

The legs are short and the general appearance of the fly is robust, black to dark brown in colour, and with white, grey or silver markings in some species. It is a strong flier and has been found many miles away from its breeding site. Only the female feeds on blood and is a diurnal (daytime) feeder.

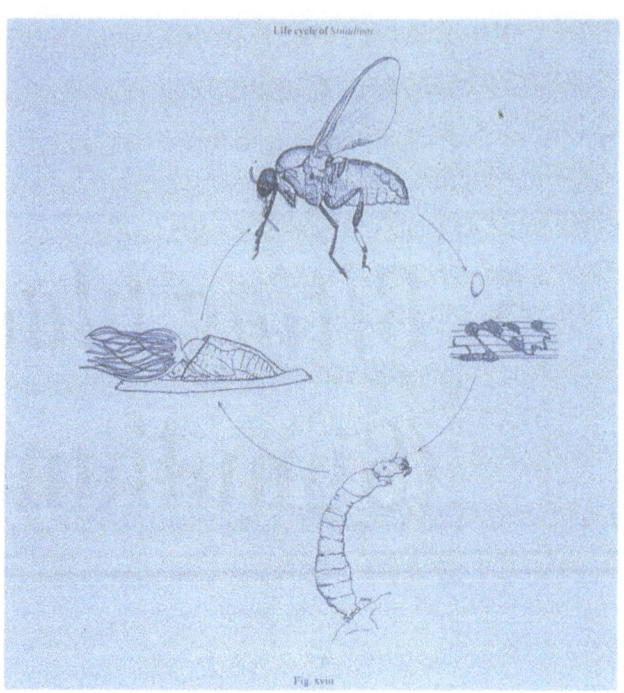

Fig. 5.3 Life cycle of *Simulium*.

LIFE CYCLE AND BREEDING SITES

The life cycle of *Simulium* is one of complete metamorphosis (Fig. 5.3). Distribution is worldwide wherever suitable conditions of well-oxygenated water exist in which the immature stages develop. Eggs are laid by the gravid female and then glued to rocks or stones, which are awash with well-oxygenated water, from slow-flowing streams to rivers in which the current is torrential. The female fly may even dive below the water surface to lay eggs on fronds of vegetation (Fig. 5.4). The eggs hatch after a few days into minute indian club-shaped larvae. The larva (Fig. 5.5) attaches itself to the substrate below the water surface by a small sucker on the thoracic proleg and by a larger anal sucker. By means of these suckers, the larva 'loops' its way through the water. One species, *Simulium naevi* in tropical Africa, attaches itself to the shell of the freshwater crab at this larval stage.

The larva feeds by filtering particles from the water and brushing microorganisms into the mouth using a pair of prominent mouth brushes (Fig. 5.6). Oxygen is obtained from the water by means of anal gills (Fig. 5.7), hence the need for well-oxygenated water. The larva undergoes six moults and reaches the pupal stage in 3 to 10 weeks. The mature larva will spin a slipper-like cocoon around itself while

Fig. 5.4 Breeding sites of *Simulium*. (a) Stream on Dartmoor, England. (b) Belize.

Fig. 5.5 *Simulium* larva.

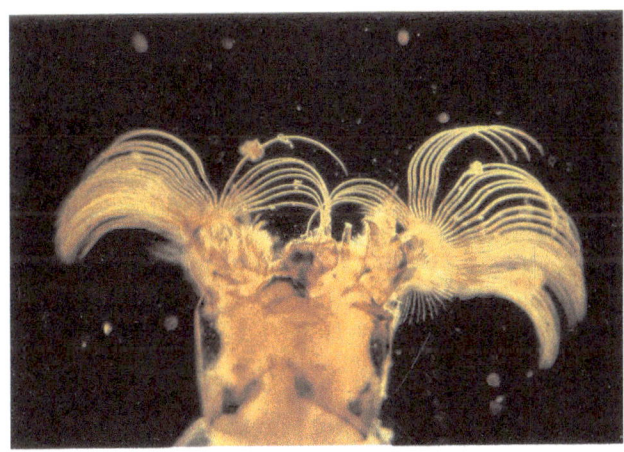

Fig. 5.6 Mouth-brushes of *Simulium* larva.

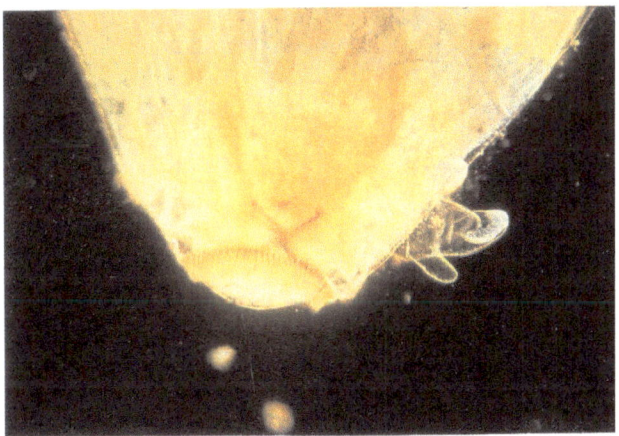

Fig. 5.7 Anal gills of *Simulium* larva.

sticking to various substrates such as rocks, stones or water plants. Inside the cocoon the larva will pupate (Fig. 5.8), extending long, paired respiratory filaments from the open end. The pupa does not move or feed. The adult emerges into the water and is carried to the surface on a bubble of air, being washed downstream at the same time. Adults emerge in late spring and early summer in cooler regions, often attacking in large numbers, particularly in tundra regions. In the tropics, the life cycle is continuous.

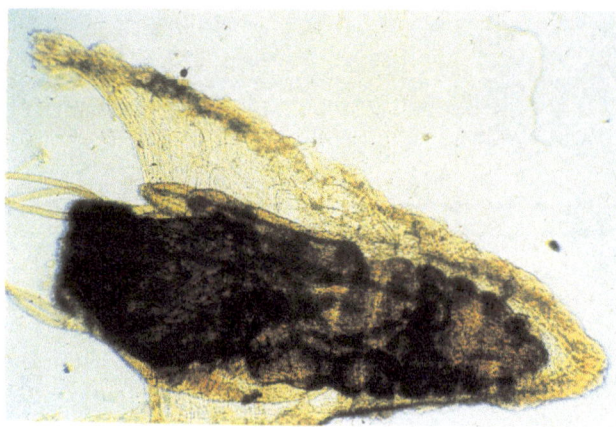

Fig. 5.8 *Simulium* pupa.

MEDICAL SIGNIFICANCE

The mouthparts of the female *Simulium* are short, broad and stubby; thus when feeding she will tend to stab the tissue and wait for the ruptured capillaries to ooze blood rather than sucking neatly like the mosquito. *Simulium* bites will typically cause considerable pain and irritation, and may frequently give rise to secondary infection. Most species will attack mammals, birds and even cold-blooded animals; a few are notorious biters of humans. Once the punctum has healed, a black scab forms (Fig. 5.9), which is typical of a *Simulium* bite.

Onchocerciasis

The blood-sucking habit of the female enables it to act as a vector of the filarial worm *Onchocerca volvulus*, causing the condition onchocerciasis in many parts of tropical Africa (carried by *Simulium damnosum* and *Simulium naevi*) and Central and South America (carried by *Simulium ochraceum*) (Fig. 5.10 (*opposite*))

In the infected human, the adult worms of *Onchocerca volvulus* form subcutaneous nodules, especially over bony prominences, which can be felt. Most of the disease processes result from an allergic reaction to microfilariae, which are released into the skin and eyes (Fig. 5.11). In the skin, this causes severe itching, papule formation and thickening and loosening of the skin, which subsequently hangs down in folds. Arthritis may also occur. The most disabling pathology occurs in the eye, where the allergic reaction to microfilariae causes sclerosing keratitis, anterior uveitis, choroiditis and optic nerve atrophy, leading to blindness ('river blindness').

Onchocerca volvulus is a tissue rather than a blood parasite, and the way in which the fly feeds enables it to take up large numbers of microfilariae. However, only two to three will develop in infective larvae in the thoracic muscles of the fly, being passed on at a subsequent blood feed about 1 week later through the thin labial membrane of the proboscis. The parasite is found in the peripheral tissue of the host during the day; this corresponds with the diurnal feeding habits of the vector.

Diagnosis is made by the discovery of microfilariae in snips of skin that have been incubated in saline. The microfilarial drug Ivermectin is used in treatment, ideally in a 6-monthly dose until the adult worms finally die, which may take up to 15 years.

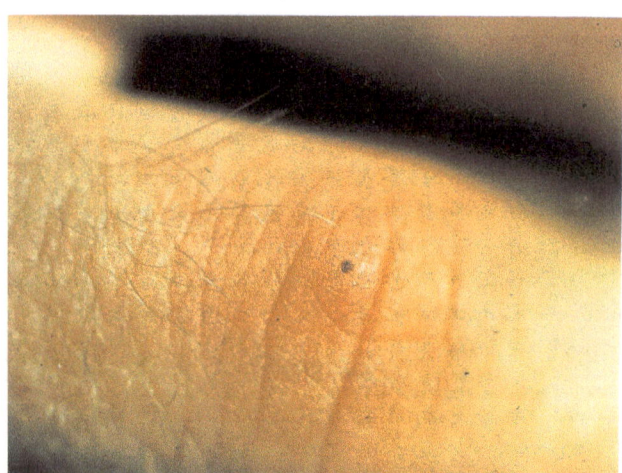

Fig. 5.9 *Simulium* bite (scab).

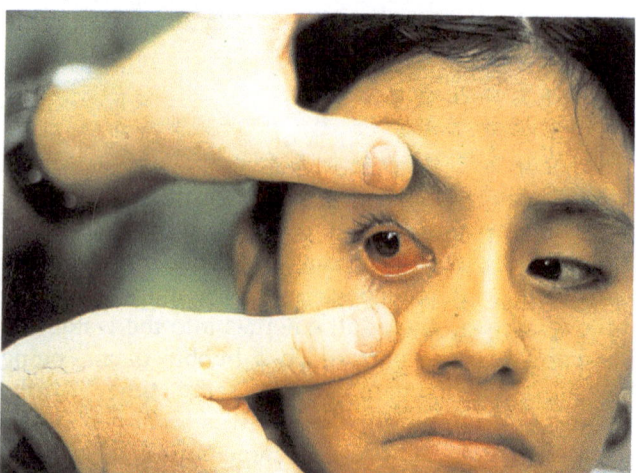

Fig. 5.11 Presence of the filarial worm causing iritis.
This may cause 'river blindness'.

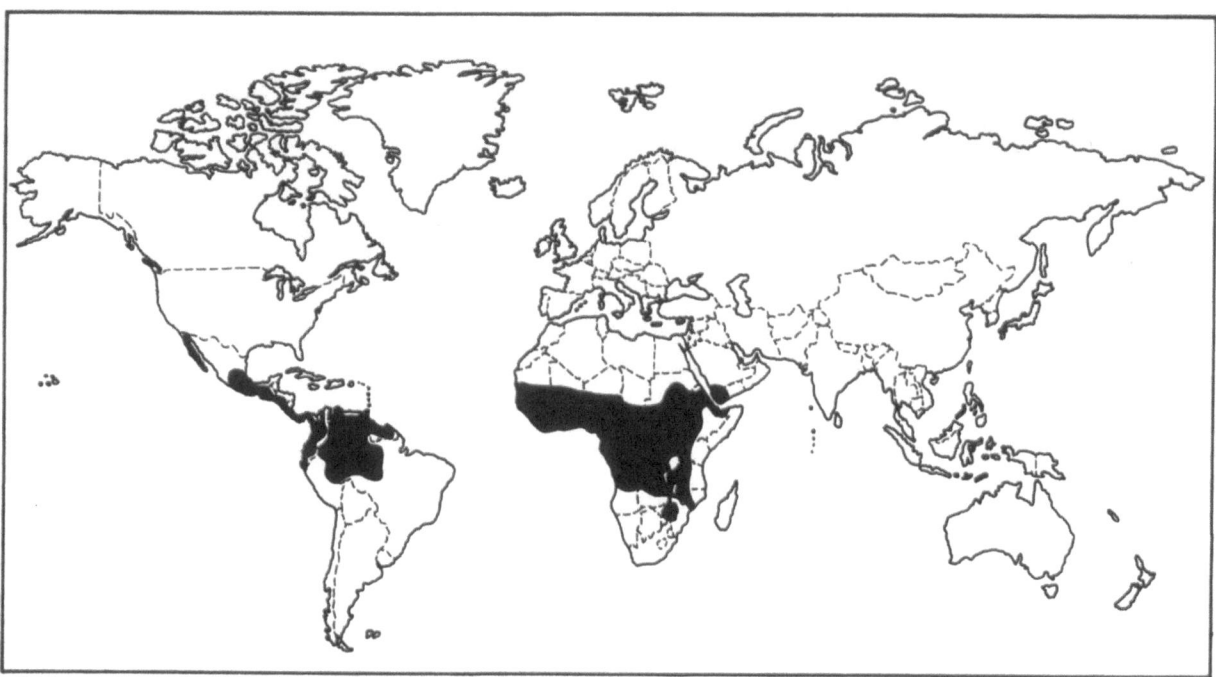

Fig. 5.10 Distribution of onchocerciasis.

6. Horseflies

INTRODUCTION AND DESCRIPTION

Horseflies make up the family Tabanidae in the suborder Brachycera. They are robust two-winged flies with antennae of three segments, the third of which has a series of annulations known as the style (Fig. 6.1). There are some 2000 species varying in length from 5 to 30 mm in several genera, the most important of which are *Tabanus*, *Haematopota* and *Chrysops*. Female Tabanids are blood-sucking (Fig. 6.2) and some species will readily transmit the filarial worm *Loa loa*, causing loaisis in humans.

Tabanids can be recognized by a combination of features: they are medium to large burly flies with three-segmented styled antennae; there is always a hexagonal cell in the centre of each wing, known as a discal cell (Fig. 6.3); and each foot has three pads known as pulvilli (Fig. 6.4). The mouthparts are broad and dagger-like, and the large compound eyes

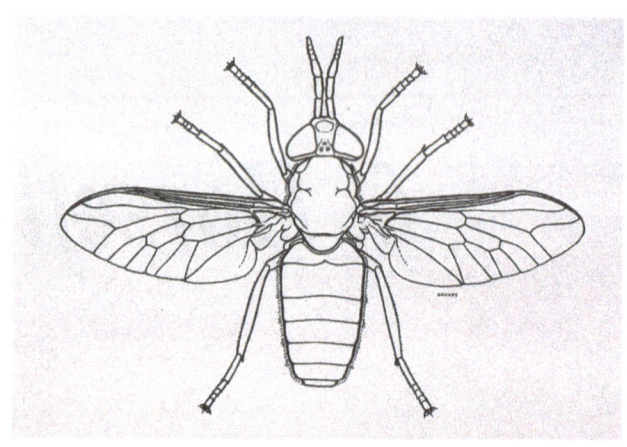

Fig. 6.3 Tabanid wing with hexagonal discal cell.

Fig. 6.1 Typical tabanid antenna. An annulated style is present on the end of the third segment.

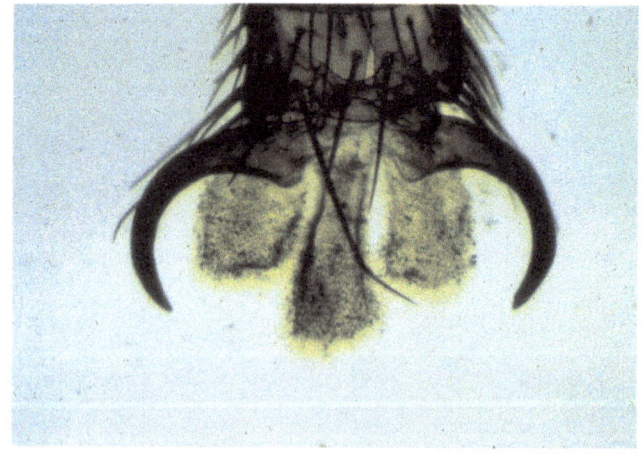

Fig. 6.4 Tabanid foot with pulvilli.

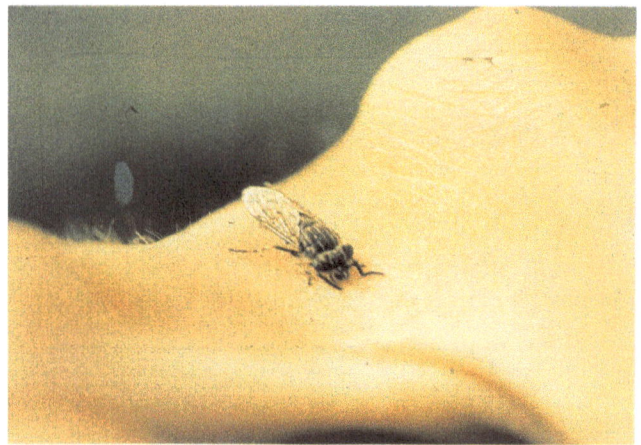

Fig. 6.2 Female tabanid sucking blood.

Fig. 6.5 Tabanid head.

are often strikingly coloured in live specimens (Fig. 6.5), being close together (holoptic) in the non-biting male but spaced apart (dichoptic) in the female.

The three important species can be easily differentiated by the appearance of the wing; *Tabanus* has clear wings (Fig. 6.6), *Haematopota* has mottled wings (Fig. 6.7) and *Chrysops* has a dark band across each wing (Fig. 6.8).

LIFE CYCLE AND BREEDING SITES

The female horsefly will lay several hundred eggs, often in two or three layers, in a wide range of wet and damp sites depending on the species. For example, some *Chrysops* will prefer the waterlogged margin of a stream or pond, while some *Haematopota* will use the damp soil or grass stems in pastureland (Fig. 6.9). From the egg a maggot-like larva emerges,

Fig. 6.6 *Tabanus* species.

Fig. 6.7 *Haematopota* species.

Fig. 6.8 *Chrysops* species.

Fig. 6.9 *Haematopota pluvialis* breeding site. In this meadow each red peg marks the position of one larva.

Fig. 6.10 Tabanid (*Haematopota pluvialis*) larva.

which is well-segmented and tapers at both ends (Fig. 6.10). The larva feeds on soft-bodied grubs and other arthropods in the environment. Development may take several months, or as long as 3 years in temperate regions where it must typically be subjected to frost. The larva will develop through as many as ten moults before crawling to higher ground to pupate (Fig. 6.11). The pupa is chrysalis-like (Fig. 6.12) and the adult emerges after only a few weeks.

Fig. 6.11 *H. pluvialis* **larva and pupa.**

Fig. 6.12 **Tabanid pupa.**

Fig. 6.13 **Horse fly bites, showing oedema.**

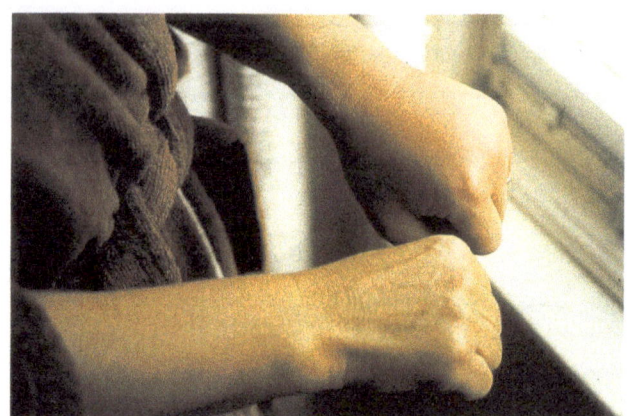

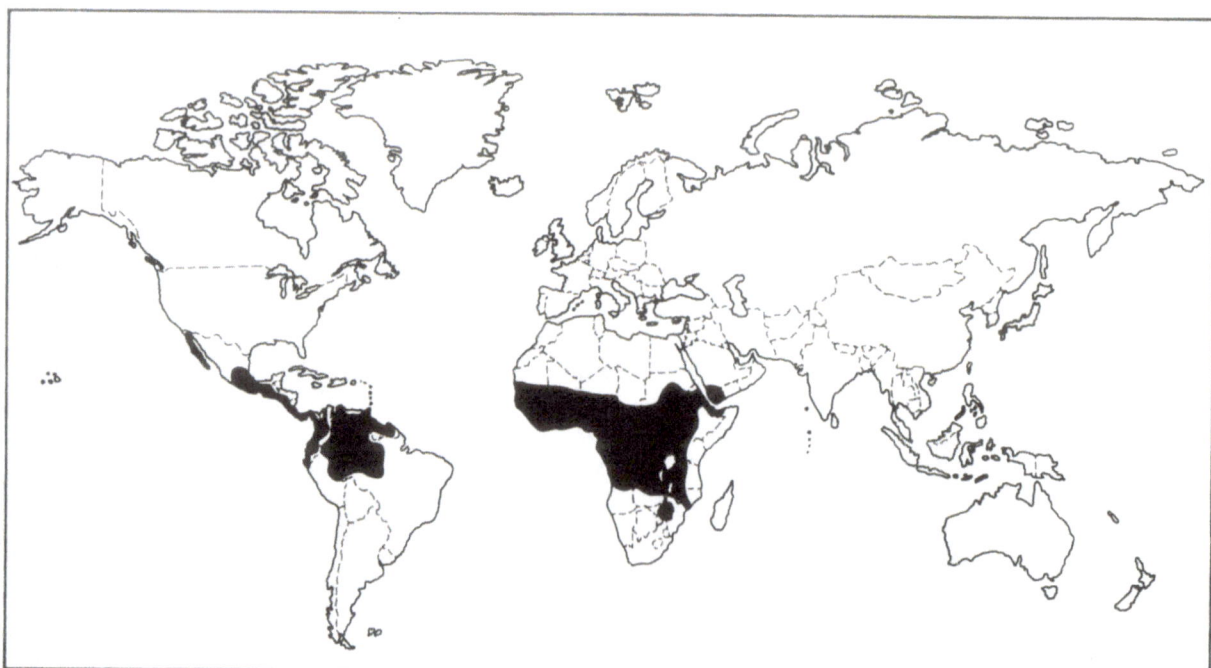

Fig. 6.14 **Distribution of loaisis.**

MEDICAL SIGNIFICANCE

The bite of the female horsefly can cause a severe reaction in a sensitized host, with considerable oedematous swelling, pain and irritation (Fig. 6.13).

Loaisis

Some species of *Chrysops*, *C. silacea* and *C. dimidiata* in tropical West Africa (Sierra Leone to Ghana and Nigeria), and *C. distinctipennis* in Central Africa (southern Sudan and Uganda), will transmit *Loa loa* to humans (Fig. 6.14). The microfilaria is taken up by the female horsefly from the peripheral blood during the day (diurnal periodicity). Development through two larval stages (Fig. 6.15) occurs in the thoracic muscles of the vector for 10 to 12 days before migrating to the proboscis and piercing the labial membrane to reach a new host at a subsequent blood feed.

Loaisis occurs only in tropical West and Central Africa. Adult worms live under the skin and migrate around the body, causing allergic ('Calabar') swellings and abscesses at the site of dead worms. The adult worm becomes visible to the patient if it crosses the anterior chamber of the eye, but it does not cause serious ocular damage. Occasionally the worms invade the brain, causing meningoencephal-itis. Marked peripheral blood eosinophilia is usual. Diagnosis is made by the presence of microfilariae in stained blood films. Treatment is with diethylcarba-mazine, and adult worms can be surgically removed.

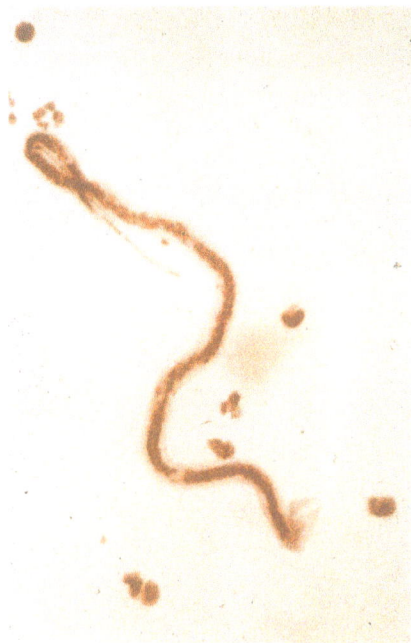

Fig. 6.15 Developing *Loa loa*.

7. The 'higher' flies

INTRODUCTION AND DESCRIPTION

Members of the suborder Athericera are sometimes referred to as the 'higher' flies (i.e. they are more highly evolved). They have short antennae comprising three segments, the two nearest to the head being very small, and the third comparatively large with a prominent bristle or arista (Fig. 7.1). Members of the Athericera vary considerably in size from 2 mm to over 10 mm and are the most specialized and numerous of the Diptera, being found in every part of the world. This suborder includes the hover flies, *Syrphidae* (Fig. 7.2), which are sometimes confused with the four-winged bees and wasps, and the fruit-flies (*Drosophila*) (Fig. 7.3) the larvae of which are often found infesting sugar-containing substances.

Most athericerous flies of medical significance are found within the group known as Calypterate flies. Among these are the bot and warble-flies (*Dermatobia*, *Gasterophilus*, *Oestrus*), and the muscoid and calliphoroid flies, for example, the common housefly (*Musca*) (Fig. 7.4), the lesser housefly (*Fannia*) (Fig. 7.5), the bluebottle (*Calliphora*) (Fig. 7.6), the green-

Fig. 7.3 *Drosophila.*

Fig. 7.1 **Calypterate antenna.**

Fig. 7.2 **Syrphid.**

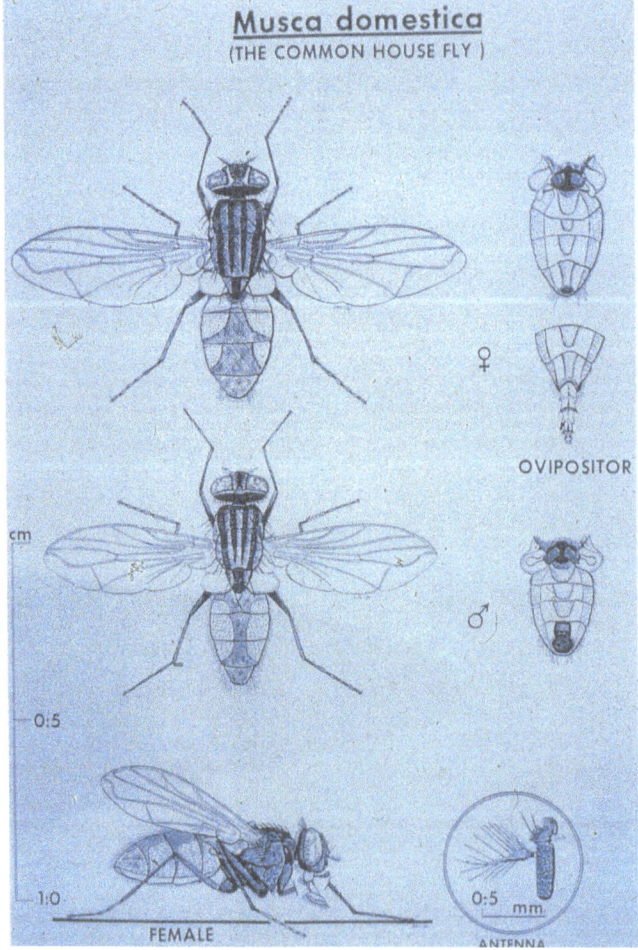

Fig. 7.4 *Musca.*

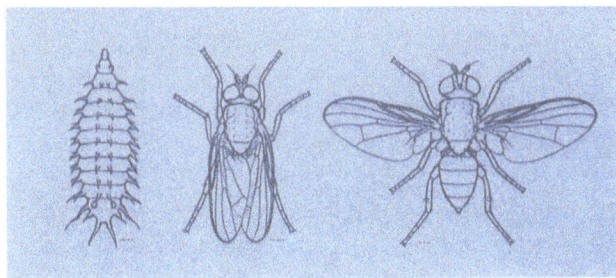

Fig. 7.5 *Fannia.*

Fig. 7.6 *Calliphora.*

bottle (*Lucilia*) (Fig 7.7), the tumbu fly (*Cordylobia*) (Fig. 7.8), the grey flesh-fly (*Sarcophaga, Wohlfahrtia*) (Fig. 7.9), and the blood-sucking tsetse fly (*Glossina*) (Fig. 7.10) and the stable fly (*Stomoxys*) (Fig. 7.11).

Calypterate flies are of medical and public health significance for several reasons. Some, for example *Musca, Fannia* and *Calliphora*, will act as mechanical vectors of pathogenic organisms, especially those causing excremental diseases due to their close association with human excrement and food consumed by humans. Others, such as *Lucilia, Sarcophaga* and *Cordylobia*, will lay eggs or sometimes larvae on soiled or damaged human tissue on which the maggots will feed and develop, causing the condition myiasis. *Glossina* and *Stomoxys* will suck human blood, and *Glossina* is also the vector of African trypanosomiasis in tropical Africa.

All calypterate flies of medical importance are medium to large, burly insects with conspicuous compound eyes, a well developed thorax and an abdomen of only four to five visible segments. The wings are broad and translucent with only six main veins, the pattern of which is often a useful guide to the genus of fly concerned.

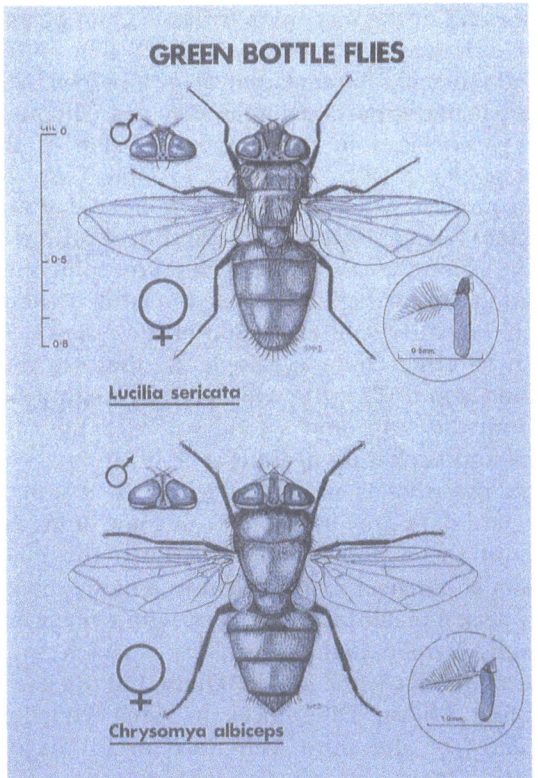

Fig. 7.7 *Lucilia* and *Chrysomya.*

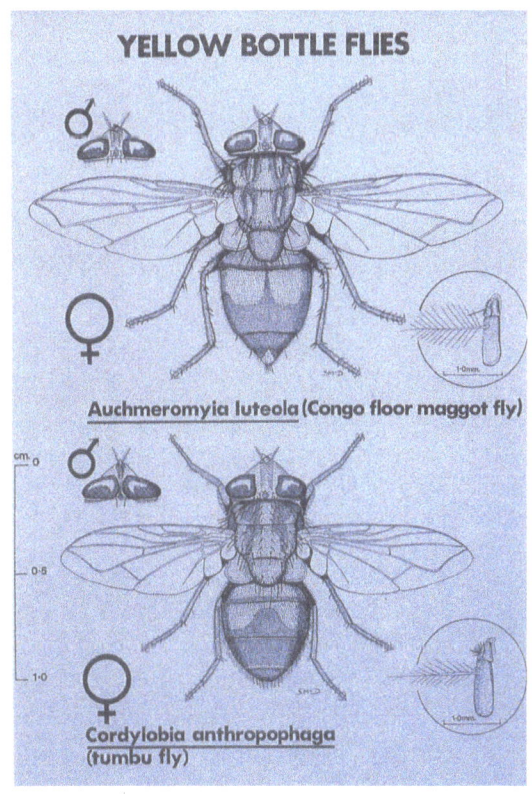

Fig. 7.8 *Cordylobia* and *Auchmeromyia.*

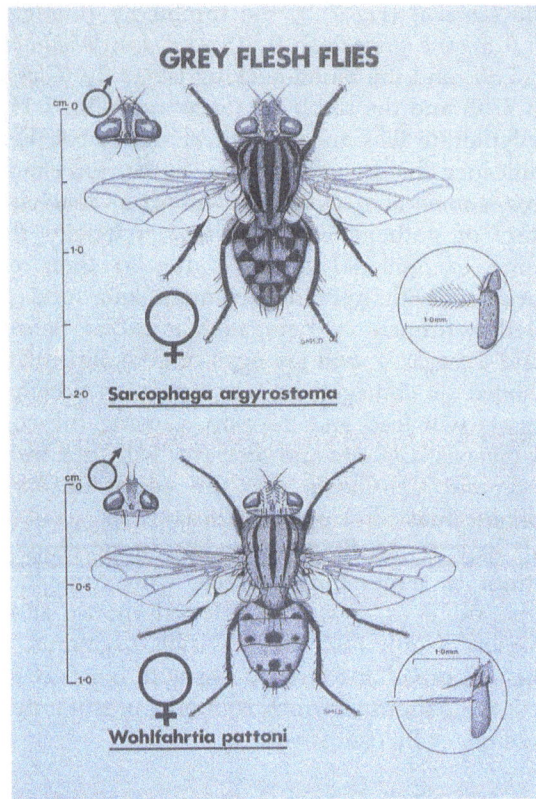

Fig. 7.9 *Sarcophaga* and *Wohlfahrtia*.

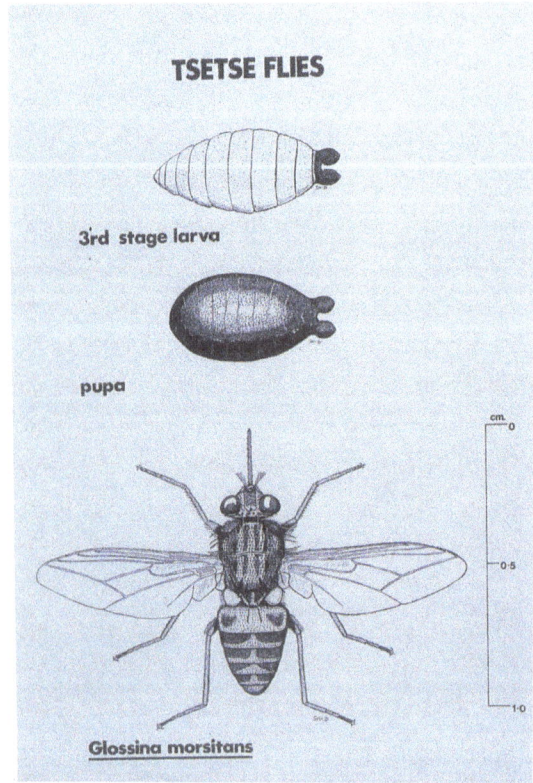

Fig. 7.10 *Glossina.*

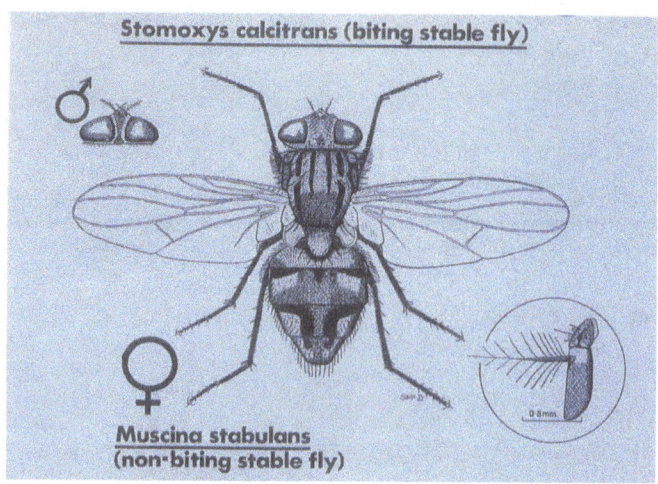

Fig. 7.11 *Stomoxys.*

LIFE CYCLE

The life cycle of a typical calypterate fly is, like all Diptera, one of complete metamorphosis (Fig. 7.12). Eggs are laid on decomposing or sound organic matter (Fig. 7.13) from which maggot-like larvae will hatch (Fig. 7.14), with a vestigial head tapering anteriorly to pointed mouth hooks (Fig. 7.15). The posterior end of the maggot is truncated and bears a pair of spiracles on a flattened plate (Fig. 7.16). There are usually three larval stages, which feed voraciously and grow considerably in size. The pupa is always enclosed in the unmoulted skin of the third-stage larva (Fig. 7.17) and is more correctly called a puparium (Fig. 7.18). It is non-feeding and completely sessile, oval in shape and dark brown or black in colour. Athericerous flies are sometimes grouped as Cyclorrhapha because the adult emerges through a circular split at the top of the pupal case. As with all other Diptera, calypterate flies can only take fluid food. The fly will typically regurgitate fluid from the crop onto a food source, liquify the material and suck it up again (Fig. 7.19).

Musca domestica is approximately 6 to 9 mm in length, and dark grey in colour with some orange or yellow on the abdomen, particularly in the male. *Calliphora* is between 5 and 7 mm in length, robust, bristly, dark metallic blue in colour, with a noisy and apparently erratic flight. *Fannia* is a much more slender fly, about 5 to 7 mm in length, but otherwise resembling *Musca*. Differences in wing venation, particularly of the third vein, are useful in differentiating these and other calypterate genera (Fig. 7.20).

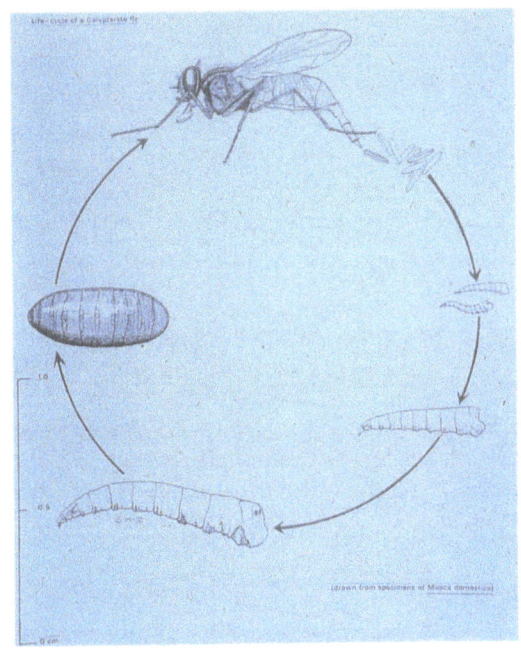

Fig. 7.12   Life cycle of *Musca domestica*.

Fig. 7.15   Larval mouthparts.

Fig. 7.13   Eggs of Calypterate fly.

(a)

(b)

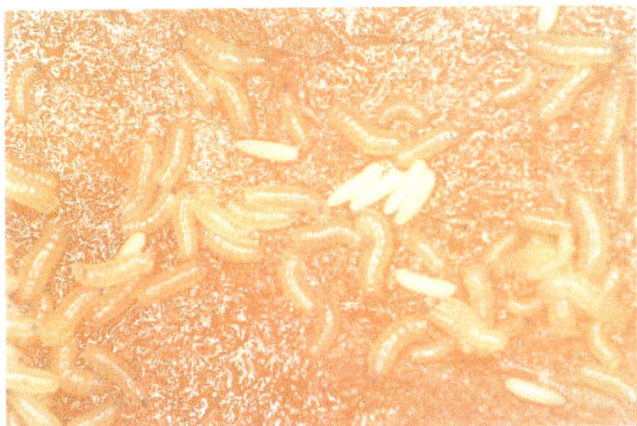

Fig. 7.14   Larvae of Calypterate fly.

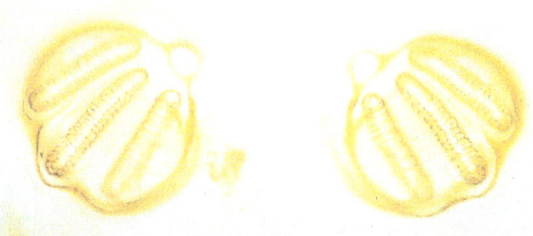

Fig. 7.16   Larval spiracles. (a) *Musca.* (b) *Calliphora.*

Fig. 7.17 Metamorphosis of the mature larva into the pupal stage (puparium).

Fig. 7.18 Puparium.

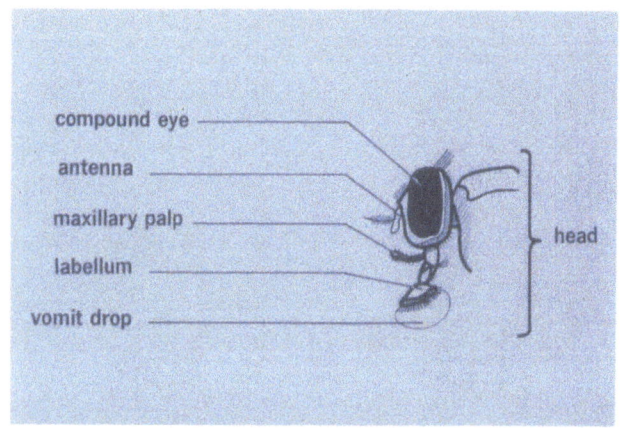

Fig. 7.19 Calypterate fly mouthparts.

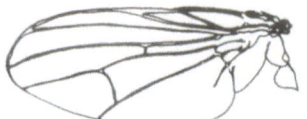

1. *Fannia canicularis* (Linnaeus, 1761)

2. *Stomoxys calcitrans* (Linnaeus, 1758)

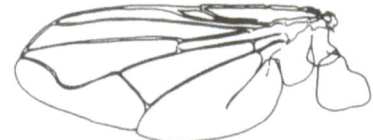

3. *Musca domestica* (Linnaeus, 1758)

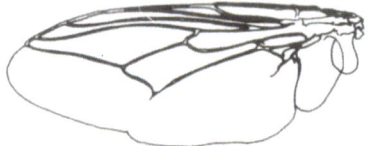

4. *Glossina palpalis* (Robineau-Desvoidy, 1830)

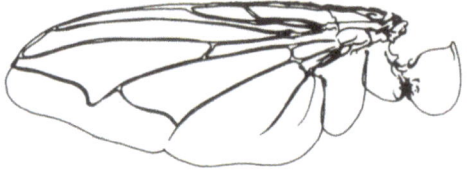

5. *Calliphora erythrocephala* (Meigen, 1826)

Note: These species cannot be identified
       by the wings alone

Fig. 7.20 Differentiation of fly genera by wing venation.

MEDICAL SIGNIFICANCE

Mechanical transmission of disease

Several genera of flies live in close association with humans and will feed readily on human faeces and other decomposing organic matter (Fig 7.21) as well as food intended for human consumption. These flies also lay eggs, several hundred at a time, on their feeding matter and in which their larval stages also develop, although the mature larva crawls to a drier environment before pupating.

When an adult fly feeds on infected decomposing matter, for example human faeces containing excremental disease organisms (Fig. 7.22), pathogens will be taken up into the gut via the mouthparts.

(a)

(b)

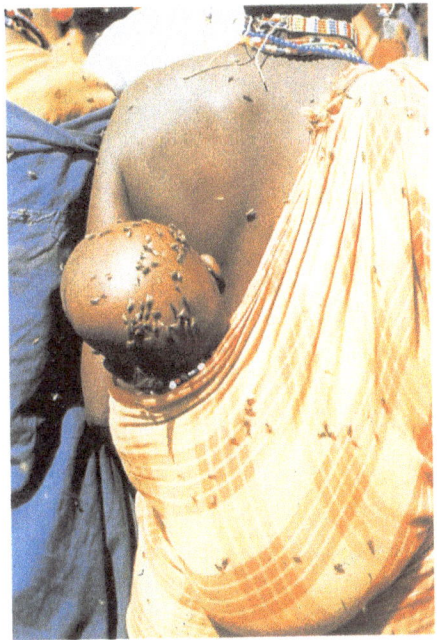
(c)

Fig. 7.21   Mechanical transmission of disease.
(a) Typical feeding site of flies in a kitchen with a pit latrine (Botswana).(b) Fly larvae and pupae, and American cockroaches were present; easy access to the kitchen was achieved via a vent pipe. (c) Flies may then settle on humans, transmitting disease.

Fig. 7.22   Flies feeding on human faeces and other decaying matter.

Organisms will stick to the hairs of the legs and body and cling to the minute bristles on the footpads as the fly walks over the faeces. Having taken a meal from this source, the insect may fly into a kitchen or dining room, land on food which is left unprotected and go through the same process of feeding. It will walk over food, cleaning the legs and body and depositing disease organisms as it moves. It will vomit fluid containing organisms from the previous meal in order to liquify its present food source, and at the same time it will defaecate the remains of previous meals of infected material. Organisms carried in this way include those causing typhoid, paratyphoid, amoebic and bacillary dysentery and cholera, as well as other forms of gastroenteritis. It may also transmit eggs of certain worms that will infect the human alimentary canal.

The flies particularly concerned with this mechanical form of disease transmission are the housefly, *Musca domestica*, the bluebottle, *Calliphora*, and the lesser housefly, *Fannia*. (These flies can be seen in Figs 7.4 and 7.5.) In temperate regions, large numbers of cluster flies (Fig. 7.23) may sometimes be found hibernating in lofts and roof spaces from where they emerge in early spring, often into the building itself, gathering onto windows and other areas.

Myiasis

Myiasis is the term given to an infestation of living tissue by dipterous larvae. Some species of fly are unable to develop through their immature stages except on living tissue. This condition is known as specific or obligatory myiasis and is caused by species limited mainly to the tropics, for example the

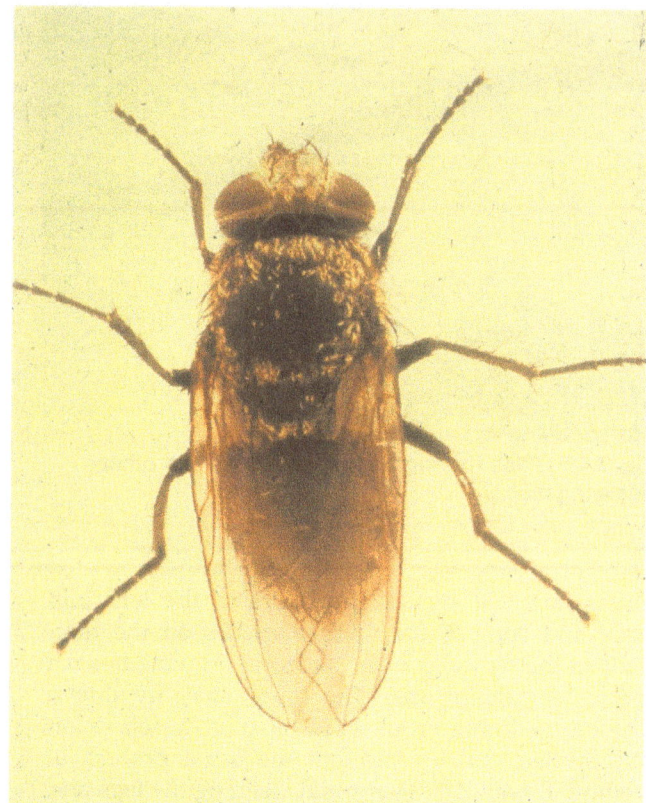

Fig. 7.23   The cluster fly *Pollenia rudis.*

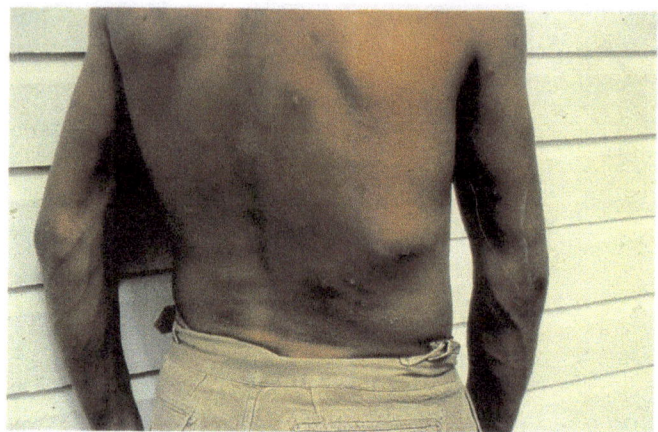

Fig. 7.25   Parasitization by the bot fly, *Dermatobia hominis* (Belize).

Fig. 7.26   Mature bot fly larva.

tumbu fly *Cordylobia anthropophaga* (Figs 7.8 and 7.24) and the screw-worm fly *Auchmeromyia luteola* (Fig. 7.8) in Africa and the bot fly *Dermatobia hominis* (Figs 7.4, 7.25 and 7.26) in Central and South America. Other species, for example the larvae of *Calliphora*, *Lucilia* (Fig. 7.27), and *Sarcophaga*, which normally

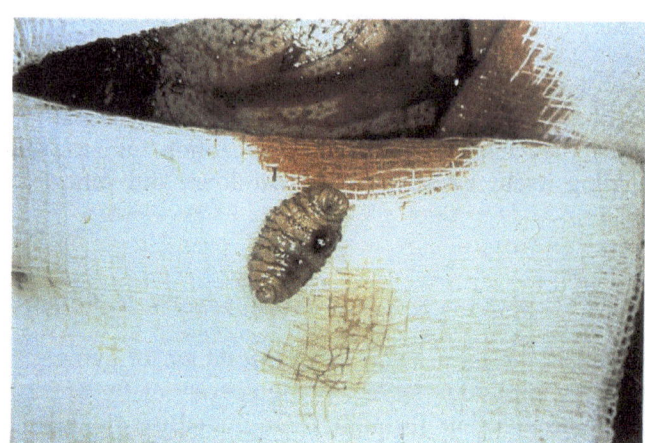

Fig. 7.24   Tumbu larva.

Fig. 7.27   Larvae of *Lucilia sericata.* This was obtained from a human axillary carcinoma (England).

develop on decaying organic matter, will readily feed on wounds or other live, damaged or contaminated tissue. This condition is known as semi-specific, wound or facultative myiasis, which sometimes occurs in the anal and genital regions. Accidental myiasis occurs when eggs or larvae are swallowed in food, resulting in intestinal disturbances. Occasionally, flies that normally parasitize animals as larvae will accidentally infest humans (Fig. 7.28).

Clinical aspects of myiasis

Cutaneous or specific myiasis occurs when the larvae of *Cordylobia anthropophaga* or *Dermatobia hominis* invade human skin. This will occur if flies have laid eggs on clothing. The infection presents as multiple painful blind boils in which secondary infection may occur. The head of the boil shows the black spiracles of the fly larva. Treatment involves suffocation of the larva with an oily ointment, after which it can be carefully grasped with forceps to ensure that the larva is removed intact.

Other fly larvae may contaminate necrotic wounds or deep ulcerating lesions caused by tumours or infections. The presence of maggots may be distressful and again, physical removal is the only method of treatment.

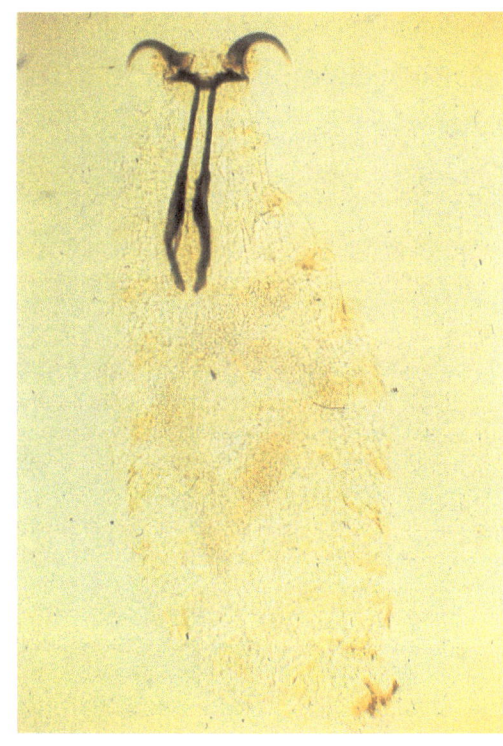

Fig. 7.28 Larva of the sheep bot or nostril fly *Oestrus ovis*. This was removed from inside the eyelid of a patient in Cyprus.

8. Tsetse flies
(Glossina)

INTRODUCTION AND DESCRIPTION

Tsetse flies belong to the genus *Glossina*, which are a unique group of insects in terms of appearance, life cycle, distribution and medical importance. There are some 23 species of tsetse, found only on the continent of Africa, south of the Sahara desert.

Species range in size from 6 to 15 mm in length, are buff brown in colour and in many respects resemble a housefly. However, there are seven visible segments on the abdomen, characteristically patterned according to species, and the wing venation is diagnostic, having a discal cell in the centre of the wing in the shape of a butcher's cleaver (Fig. 8.1). The wings are folded over each other like the blades of a pair of scissors when at rest (Fig. 8.2). Unlike most groups of blood-sucking flies, both male and female tsetse will feed exclusively on blood and have a forwardly projecting proboscis held between two palps, which remain pointing forwards during feeding. The arista, the long bristle on each antennae, is feathered (Fig. 8.3), a feature found only in this genus.

Fig. 8.3   Plumed arista of antenna.

Fig. 8.1   Tsetse fly (*Glossina* spp.). This shows the general appearance and colour, wing venation and abdominal segmentation of the fly.

Fig. 8.2   Tsetse fly. At rest, the wings are folded over the abdomen.

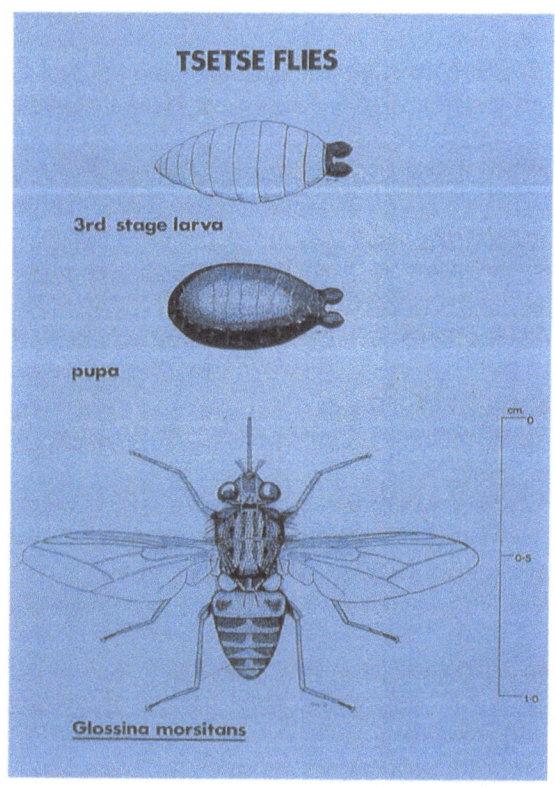

Fig. 8.4   Life cycle of *Glossina*.

LIFE CYCLE

The life cycle of the tsetse (Fig. 8.4) is a unique exception to that of the typical calypterate fly in that only a single egg at a time is fertilized as it passes down the oviduct, from sperm held in the spermatheca after mating. This egg hatches in the uterine pouch, and the resulting larva (Fig. 8.5) will feed and develop there until fully mature. The larval mouthparts are attached to a gland in the wall of the uterus, and as the larva develops oxygen is obtained through spiracles on the peripneustic lobes, which are extruded through the genital orifice of the female. Only when the larva is fully grown is it 'laid', pupating immediately without further feeding (Fig. 8.6). Larval development takes about 2 weeks and the pupa will hatch in a further 2 to 3 weeks (Fig. 8.7). By continually filling and emptying the

Fig. 8.6 *Glossina* **larva and pupae.**

Fig. 8.5 *Glossina* **larva with peripneustic lobes.**

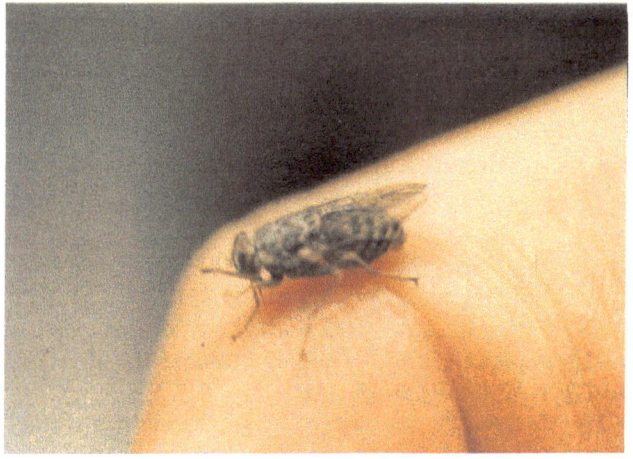

Fig. 8.7 **Adult tsetse fly feeding.**

ptilinal bladder with body fluid, the fly will remove the cap of the pupal case and dig its way out of the substrate. The female fly will lay only six to 12 larvae during a lifetime, placing them carefully in the substrate, away from predators.

MEDICAL SIGNIFICANCE

Much of the early work on the epidemiology of African trypanosomiasis was carried out by the Uganda Sleeping Sickness Commission, which reported in 1897.

Two groups of *Glossina* are involved in the transmission of African trypanosomiasis or sleeping sickness to humans (Fig. 8.8). In addition, other species

transmit the trypanosome causing ngana in cattle. The riverine areas of West Africa, based mainly on the Niger and Congo basins, are the habitat of *Glossina palpalis*, vector of the chronic Gambian form of the disease caused by *Trypanosoma brucei gambiense* (Fig. 8.9). In the hinterland where the rivers dry into pools in the hot season, the vector is the smaller species *G. tachinoides*. Both species will feed most readily on humans although, unusually, *G. palpalis* is not averse to reptilian blood.

In the game areas of East Africa, with its wide plains and bush country, species of tsetse occur that do not require the humid conditions of river bank vegetation. *Glossina morsitans* is found in wooded savanna where rocks and trees provide the shade that is required (Fig. 8.10). *G. swynnertoni* is capable

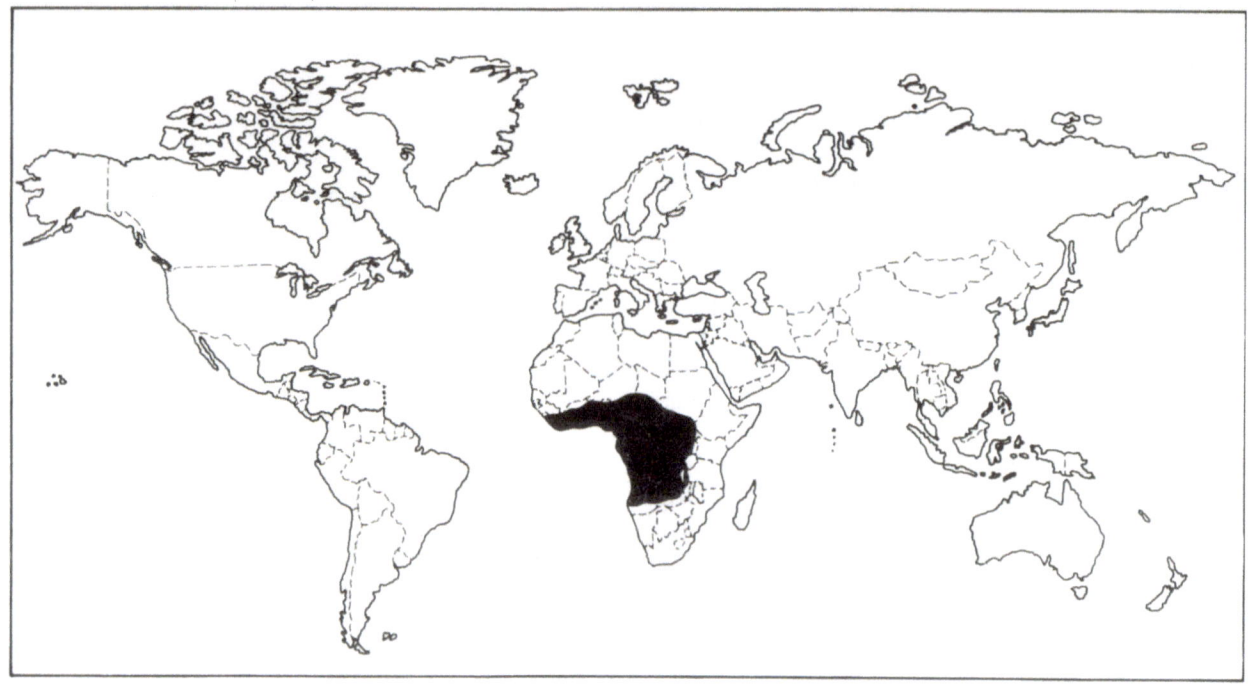

▨	Trypanosoma Rhodesiense
▨	Trypanosoma Garnbiense

Fig. 8.8   Distribution of African trypanosomiasis.

of surviving in much more open terrain where the only shade is provided by the host animal itself. Both species are associated with the Rhodesian form of trypanosome found in wild game and infective to humans, *T. brucei rhodesiense*.

The tsetse fly acquires trypanosomes from the peripheral blood of the animal or human host during a blood meal. The organisms enlarge and multiply in the alimentary canal of the fly, producing a large number of slender forms. These proceed into the lumen of the gut inside the peritrophic membrane and then forward between the membrane and gut wall, back into the oesophagus, and thence into the salivary glands of the fly. Metamorphosis occurs during this journey, first into the crithidial form and finally into the stumpy metacyclic form, which is

Fig. 8.9   Typical riverine habitat of *Glossina palpalis*.

Fig. 8.10   Typical savanna habitat of *Glossina moristans*.

injected via the saliva into a new host. Further multiplication and development occurs in the host (Fig. 8.11).

Clinical aspects

In humans, African trypanosomiasis occurs in two forms:

1. Rhodesian (*T. brucei rhodesiense*). Fever starts within 1 to 2 weeks of the infective bite, which may be apparent as a painless blind boil or chancre. Irregular fever continues, accompanied by skin rashes, enlargement of the liver and lymph nodes, myocarditis and finally progressive meningoencephalitis leading to coma and death.

2. Gambian (*T. brucei gambiense*). The illness is more chronic and insidious, and may last several years. There is low-grade fever and some lymph node enlargement, especially in the neck. The 'sleeping sickness' stage is marked by mental deterioration, a shuffling gait, somnolence and increasing inactivity, with death finally ensuing through secondary infection or malnutrition.

Diagnosis of Rhodesian trypanosomiasis is made by the discovery of trypanosomes in the blood, and in Gambian trypanosomiasis by the discovery of trypanosomes in lymph gland fluid or cerebrospinal fluid (Fig. 8.12). Serological methods are also used. The traditional treatment drugs, Suramin for early cases and Melarsropol if the brain is involved, may be replaced by Eflornithine (DFMO).

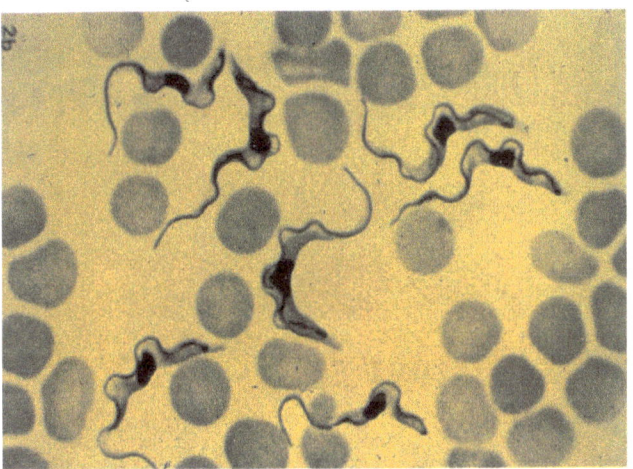

Fig. 8.11 Trypanosomes in blood smear.

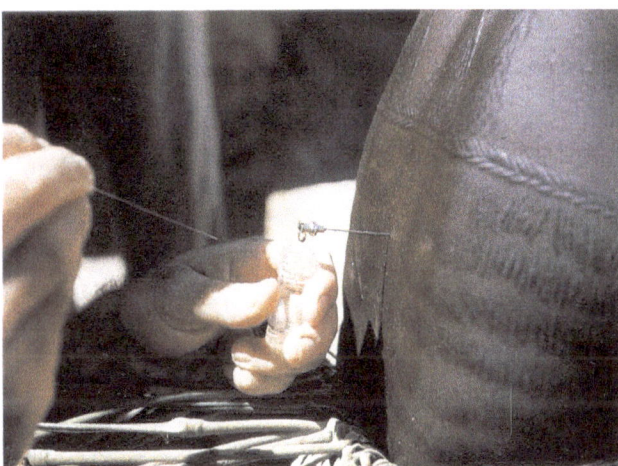

Fig. 8.12 Diagnosis of Gambian trypanosomiasis. A lumbar puncture is performed to obtain spinal fluid.

9. Fleas

INTRODUCTION AND DESCRIPTION

There are some 3000 species of fleas worldwide, making up the insectan order Siphonaptera. All are blood-sucking temporary ectoparasites of warm-blooded animals, mainly mammals and some birds. Although they are comparatively host-specific, a few species will feed readily on humans if their preferred host is unavailable. Blood is required by both sexes for nutrition and the female requires blood of the specific host in order to lay viable eggs.

Fleas (Fig. 9.1) are oval in shape and light to dark brown in colour, ranging in length from 1 to 8 mm

depending on the species. They are laterally flattened; this is a useful adaptation that enables the flea to move easily through the hairs or feathers of its host. The small head has a pair of simple eyes, although species that parasitize subterranean hosts may be blind. Short, paired antennae are held in grooves on the head and the conspicuous blood-sucking mouthparts project downwards (Fig. 9.2).

Fleas are secondarily wingless, having lost their wings during the course of evolution as ectoparasites. However, the legs (particularly the hind pair) are highly developed for jumping (Fig. 9.3) and enable the insect to leap 10 to 15 cm, a considerable distance for its size. The abdomen is bulky and obviously segmented; the coiled genitalia are usually apparent through the integument in the male (Fig. 9.4), whereas the female abdomen is more rounded and only the spermatheca is apparent (Fig. 9.5). A row of dark bristles, the combs, may be present on either side of the head (genal combs) and the thorax

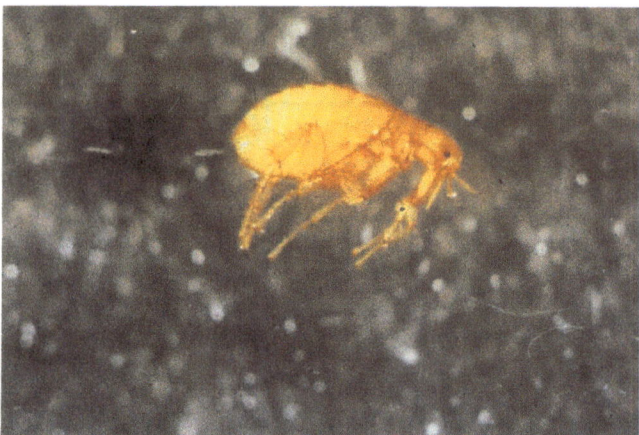

Fig. 9.1 Typical flea.

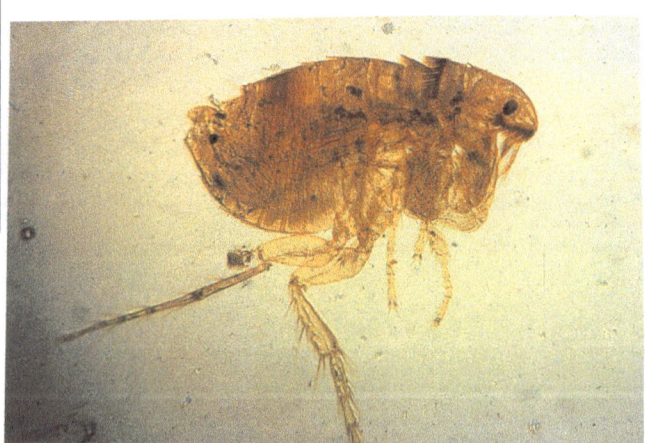

Fig. 9.3 Highly adapted legs of the flea.

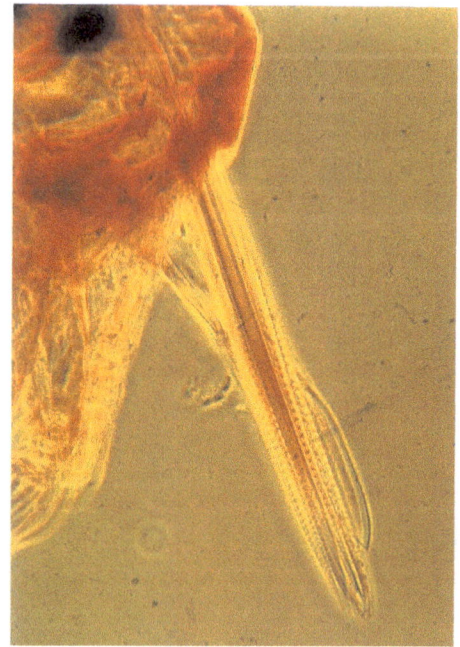

Fig. 9.2 Flea proboscis. The paired maxillary blades surround the stylet-like internal mouthparts.

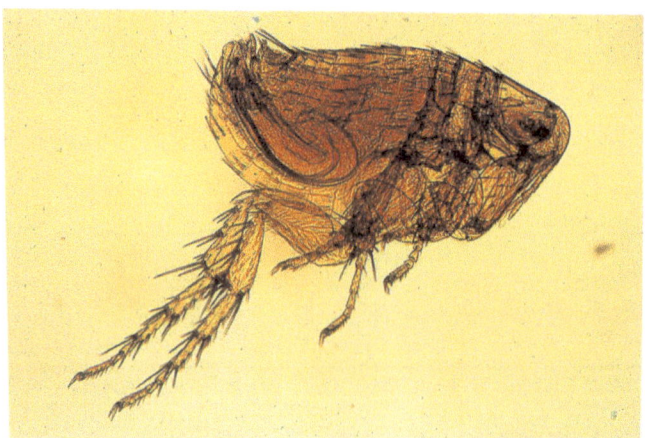

Fig. 9.4 Coiled penis and testes of the male flea. These are apparent through the exoskeleton.

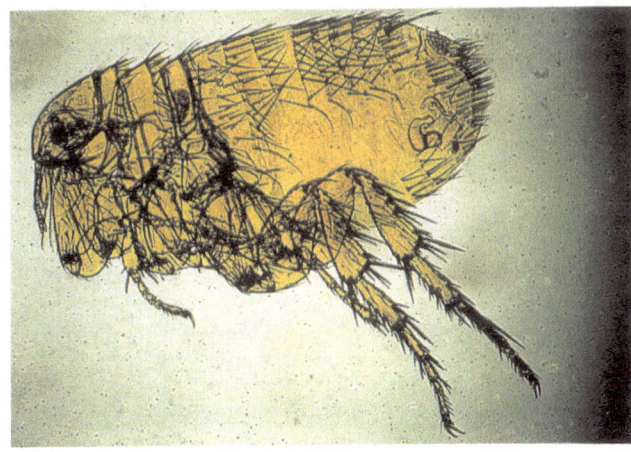

Fig. 9.5 Spermatheca in the female abdomen. The shape is often characteristic of the species.

Fig. 9.6 Genal and thoracic combs. Their presence and appearance are useful in species differentiation.

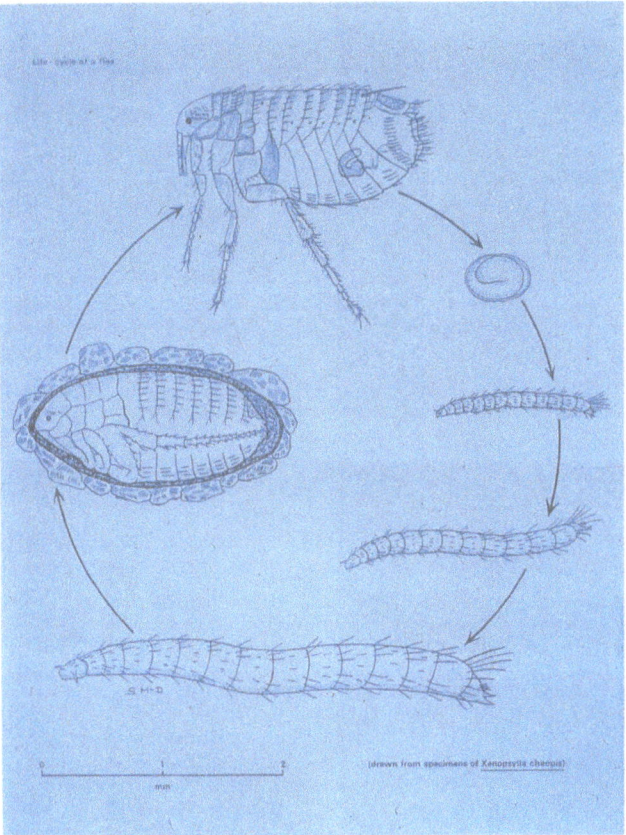

Fig. 9.7 Typical life cycle of the flea.

(Fig. 9.6); these are often useful in differentiating species.

LIFE CYCLE

The life cycle is one of complete metamorphosis (Fig. 9.7) and typically takes place in the resting place (nest or burrow) of the host. The gravid female (Fig. 9.8) will lay some 20 to 30 eggs a day in this habitat, or the eggs will sometimes be stuck lightly to the host. The eggs are 0.5 mm long, oval and pearly white (Fig. 9.9). The female flea may live for several months.

The egg hatches in about 1 week and the maggot-like larva (Fig. 9.10) emerges legless and covered in long body hairs, with two small processes (anal struts) at the posterior end. It moves rapidly in search of food, which may be in the form of organic

Fig. 9.8 Gravid female flea.

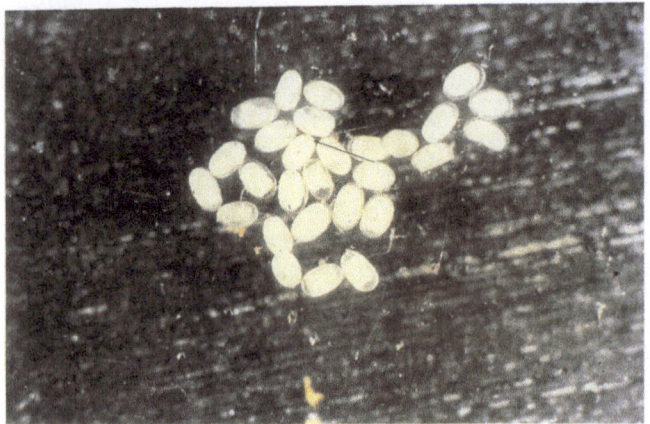

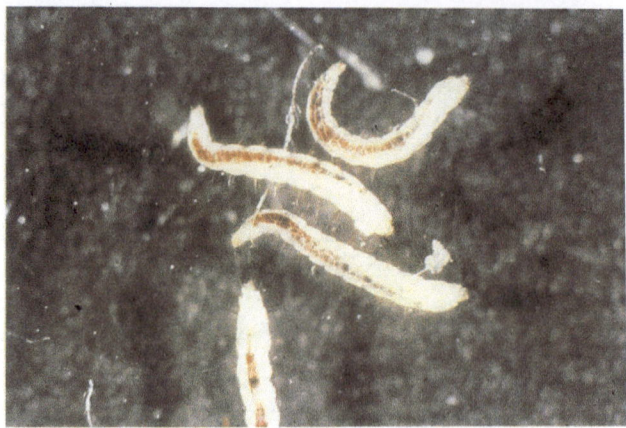

Fig. 9.11 Flea larvae. The gut contains semi-digested blood provided by the adult flea.

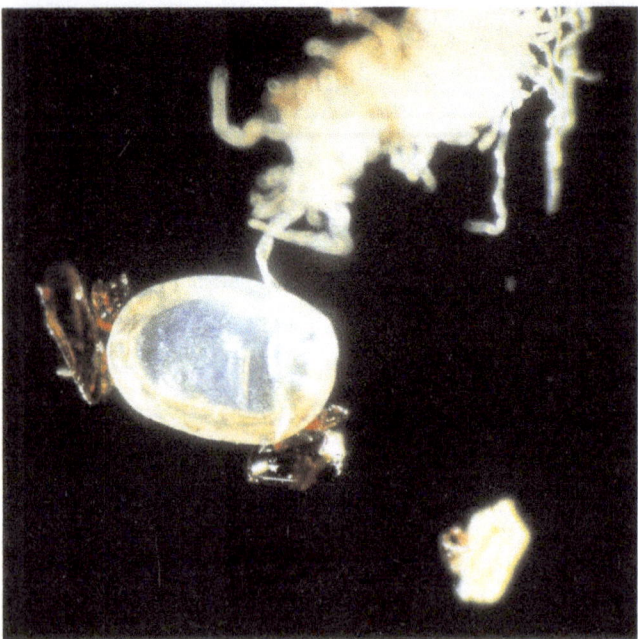

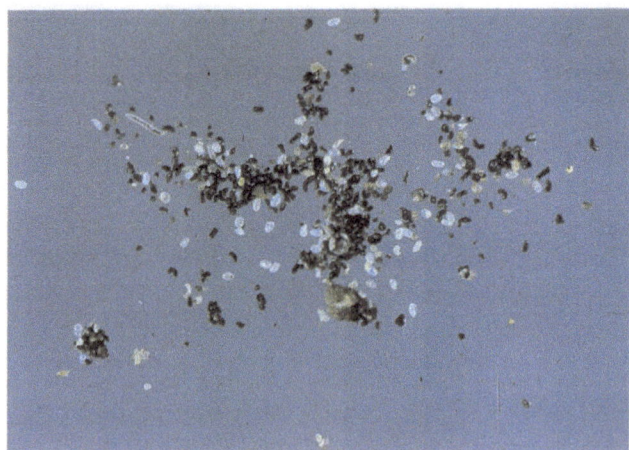

Fig. 9.12 Detritus from a cat basket showing particles of semi-digested blood from the host. This has been defaecated by the adult flea.

Fig. 9.9 Flea eggs.

Fig. 9.10 Flea larvae. Long body hairs and anal struts can be seen.

debris in the habitat of the host, but also dried, semi-digested blood excreted by the adult flea. This is often apparent in the larval gut (Fig. 9.11) and can be detected by low-power microscopic examination of the host's habitat (Fig. 9.12). After four stages or instars lasting 1 to 3 weeks, the mature larva spins a cocoon around itself inside which it pupates. Debris from the environment will stick to the outside of the cocoon (Fig. 9.13).

The adult flea emerges from the pupal case after 1 to 2 weeks but remains in the cocoon until it senses the presence of a potential blood meal in order to ensure its survival. It may remain protected in this way for many months, thus a heavy flea infestation may suddenly become apparent in a room or building that has been uninhabited for some time.

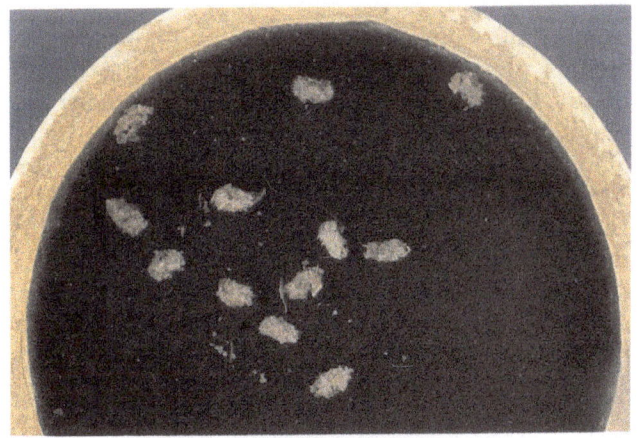

(a)

Fig. 9.14   Flea feeding on human host.

(b)

Fig. 9.13   **Flea cocoon.** (a) Intact cocoon. (b) Flea from cocoon.

Fig. 9.15   **Genal and thoracic combs of** *Ctenocephalides felis* **(cat flea).**

MEDICAL SIGNIFICANCE

Bites

Both male and female fleas in several species will feed on humans, causing irritation, redness and swelling in a sensitized host. The flea will feed by crouching low (Fig. 9.14) and inserting the proboscis into the tissue with a sawing action, injecting an anticoagulant saliva into the wound. The development of sensitivity and later immunity is similar to that encountered with mosquito bites.

In many parts of the world the so-called human flea *Pulex irritans* is still common but, in developed countries with improved housing conditions, central heating, carpeted floors, and domestic cats, the cat flea *Ctenocephalides felis* (Fig. 9.15) is the most common cause of flea bites in humans. This species will bite on the most accessible parts of the body, thus bites on the lower limb are typical (Fig. 9.16). The

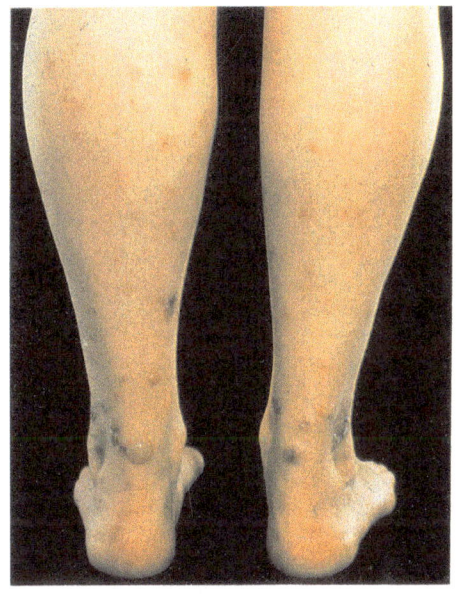

Fig. 9.16   **Typical flea bites on the lower limb.**

flea will sometimes stand on the edge of clothing (e.g. sock, underwear) and bite where the skin emerges. Thus a line of bites may be diagnostic (Fig. 9.17).

The presence of fleas in a domestic situation may be confirmed by taking brushings from pet animals onto paper (not plastic, as it makes examination more difficult) and looking for the characteristic coils or particles of dark dried blood. Debris from chairs and carpets can be examined in the same way. Sometimes eggs, larvae and cast skins may also be seen (Fig. 9.18).

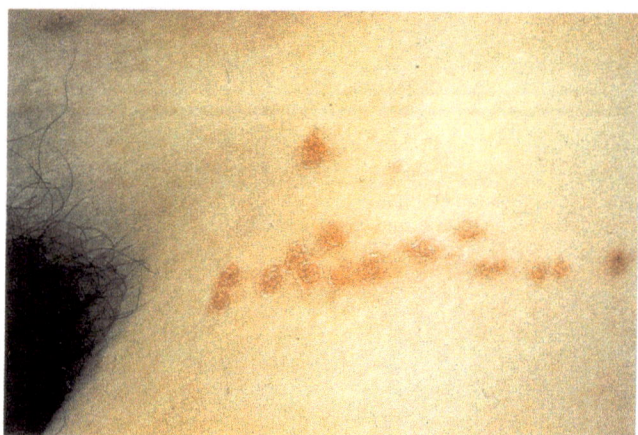

Fig. 9.17   Line of flea bites.

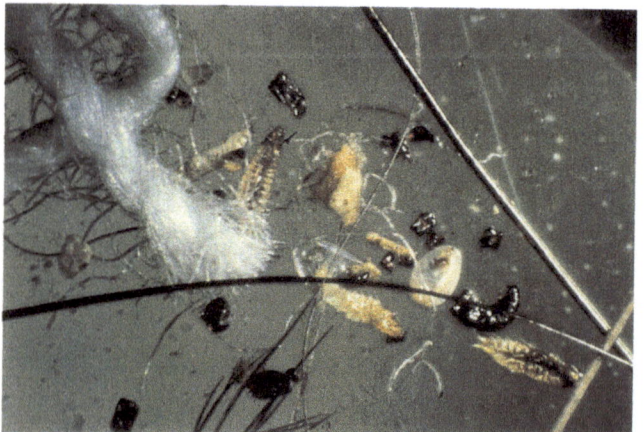

Fig. 9.18   Debris from armchair on which a cat sleeps.
This includes cat hair, cast larval skins and dried blood defaecated by the adult flea.

Plague

The plague organism, *Yersinia pestis* is found in a wide range of wild rodents in many parts of the

tropics and subtropics, and is passed from one to another by their specific fleas, with little or no detrimental effect to the host animal. If humans are bitten by an infected flea from one of these wild animals, a sylvatic infection of plague may be contracted, a condition that occurs, for example, among hunters, lumberjacks and other backwoodsmen. The organism may be carried to an urban situation, perhaps via infected wild rodents or humans, the domestic rat (common, sewer or brown rat, *Rattus norvegicus*; ship, black or roof rat *Rattus rattus*) via their own or other species of flea. The plague bacillus will usually kill the rat leaving the fleas to search for blood from the nearest available host, which is often human. *Nosopsyllus fasciatus* (Fig. 9.19a) is the rat flea of temperate regions, and

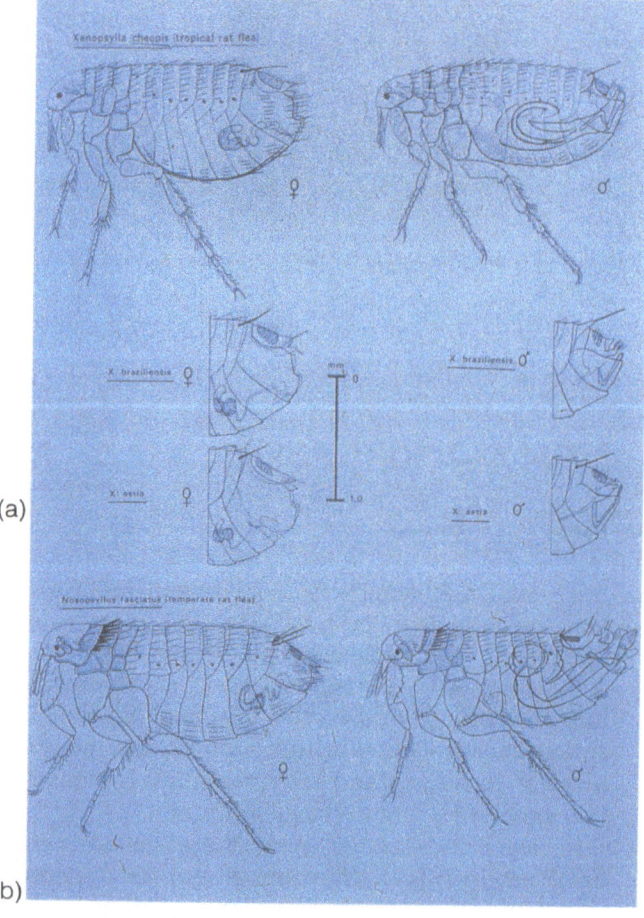

Fig. 9.19 (a) *Xenopsylla* **species of the tropical rat flea.** This shows diagnostic differences in the shape of the spermatheca in three important plague vector species. (b) *Nosopsyllus fasciatus.* This is the probable vector of pandemics such as the Black Death of medieval times.

species of *Xenopsylla* such as *X. astia*, *X. braziliensis* and *X. cheopis* (Fig. 9.19b) are found in the tropics. The bacillus will be taken up in a blood meal from an infected animal and will multiply in the proventriculus of the flea vector (Fig. 9.20), eventually blocking the alimentary canal. The flea becomes increasingly hungry, feeding frequently and regurgitating plugs of bacilli into new hosts. The flea has long been recognized as a savage blood-sucker, and has been associated with rats and the plague.

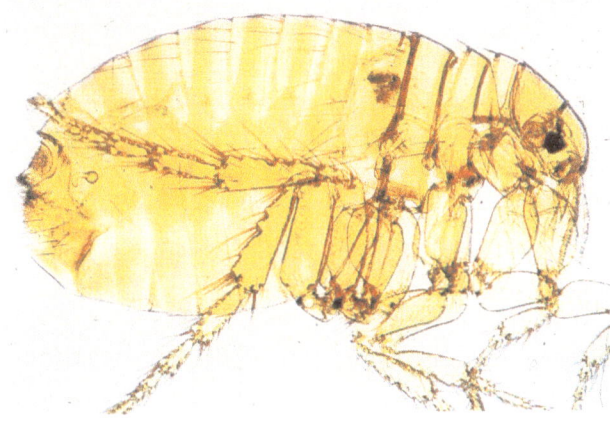

Fig. 9.20 Female *Xenopsylla cheopsis*. The proventricular valve is blocked with multiplying bacilli.

Bubonic plague causes a painful enlargement of the regional lymph node ('bubo') in the area bitten by the infected flea, accompanied by fever and, in severe cases, septicaemia with severe debilitation and delirium. In epidemics of plague, direct person-to-person transmission results from contact with pus from discharging buboes, and inhalation of material produced by patients who have developed plague pneumonia (which may be rapidly fatal). Large volumes of watery, blood-tinged and highly infective sputum are characteristic. Pneumonic plague can also result from inhalation of dust containing dried faeces from infected rats. Diagnosis is by demonstration of *Yersinia pestis* in lymph node fluid, blood or sputum. Treatment is with streptomycin, tetracycline or chloramphenicol. A preventive vaccine is available.

Murine (endemic) typhus

Several species, particularly rat fleas, will act as vectors of *Rickettsia mooseri* causing endemic or murine typhus in humans. The organism is taken up in a blood meal by the flea from an infected host, multiplies in the flea gut, and is passed to a new host via the faeces or crushed body of the flea, through abrasions in the host's skin. There are several other forms of typhus caused by different rickettsial organisms transmitted by body lice, hard ticks and *Trombicula* mites. All forms produce a feverish illness accompanied by a cough, headache, mental dullness and constipation, and often by a macular rash on the trunk. Vaccines have been used in the past but are not generally available nowadays. Diagnosis is by serology, and all forms respond to treatment with tetracyclines or chloramphenicol.

Other diseases

Humans may become infected with the infective larval stage of the tapeworms *Dipylidium caninum* and *Hymenolepis diminuta*, if infected fleas are swallowed. Fleas may also be involved in the transmission of tularaemia.

Tungiasis

A few species of the 'stick-tight' fleas will remain associated with their host for long periods. One such species is the jigger flea *Tunga penetrans*, found originally in tropical Central and South America, and lateer in tropical Africa (Fig. 9.21) and western India. The jigger flea is only about 1 mm long, with a pointed head and a telescoped thoracic region. The male abdomen is pointed (Fig. 9.22), whereas that of the female is rounded (Fig. 9.23). Both sexes feed on blood but the female will remain embedded in the tissue, particularly on the feet and hands, where she will swell with the developing eggs in the ovaries causing the characteristic 'chigger' (not to be confused with 'chigoe', which is the parasitic larva of the scrub typhus vector and reservoir *Trombicula*). Mature eggs are produced (Fig. 9.24) at a rate of up to 200 per day for 1 week to 10 days, and drop to the ground to hatch. The female dies *in situ* and may cause considerable local reaction and secondary infection.

'Chiggers' are painful swellings, often beneath a toenail (Fig. 9.25), caused by the presence of the gravid female *Tunga penetrans*. The infection is seen in tropical Central and South America, Africa and western India. Needle dissection to remove the flea is the only treatment of this condition. Subsequent disinfection is also required, as occasionally the wound can act as an entry portal for the spores of *Clostridium tetani*.

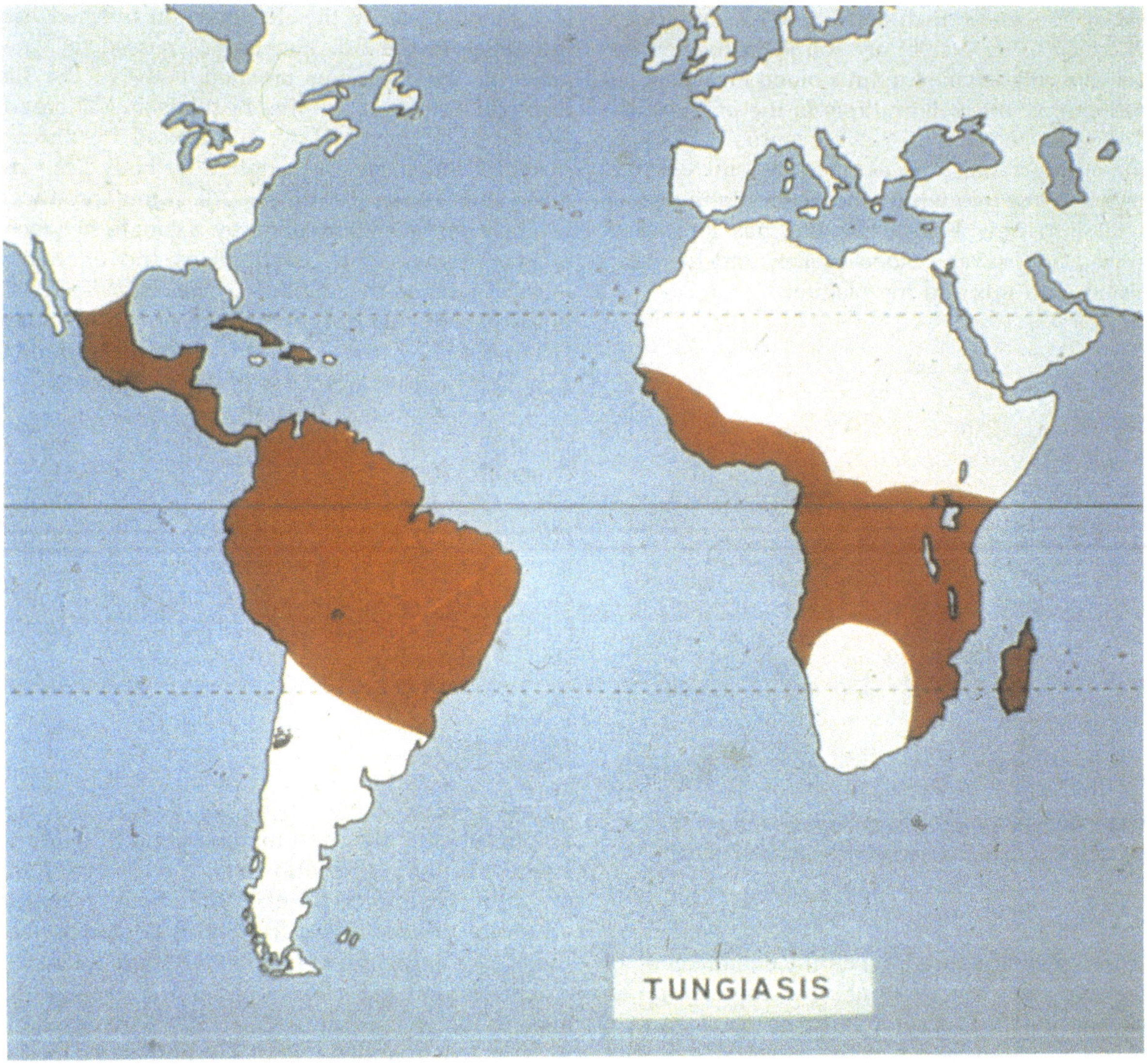

Fig. 9.21 **Distribution of tungiasis.**

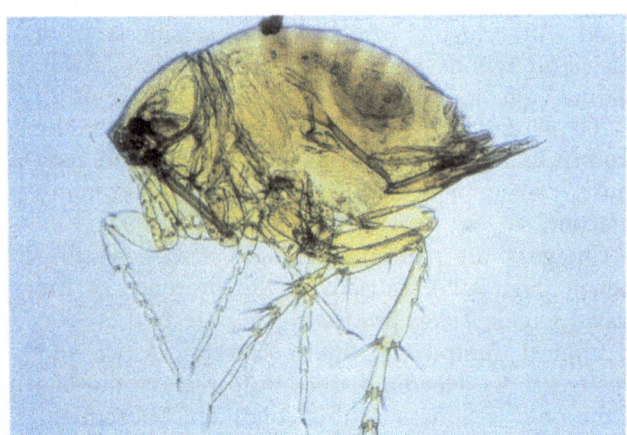

Fig. 9.22 **Male jigger flea *Tunga penetrans*.** The pointed abdomen is apparent.

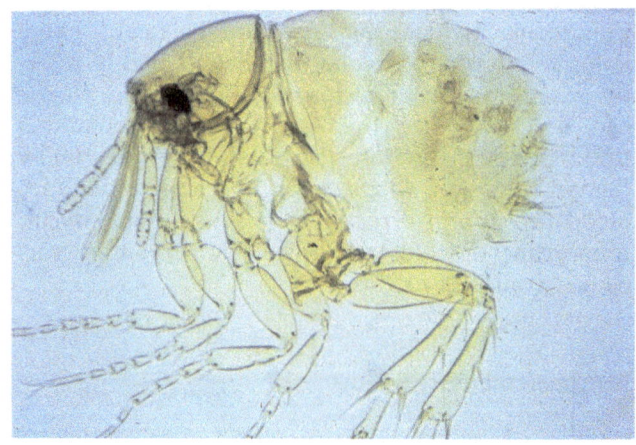

Fig. 9.23 **Female *Tunga penetrans*.** The spermatheca in the rounded abdomen can be seen.

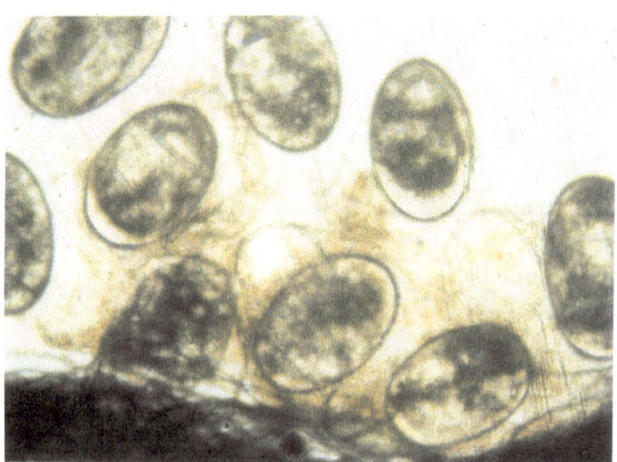

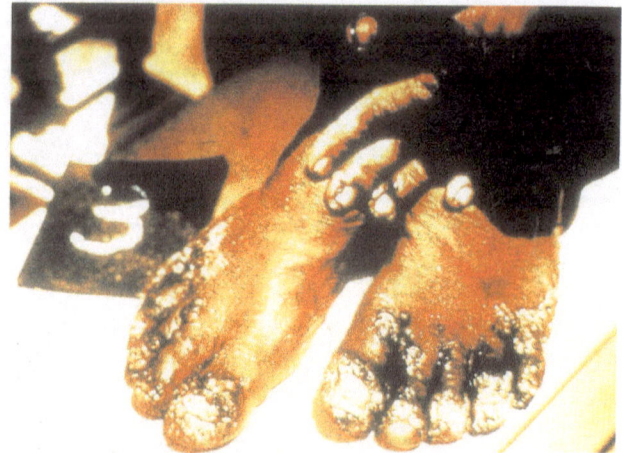

Fig. 9.24 Mature eggs of *Tunga penetrans*. Eggs are laid by the female *in situ*.

Fig. 9.25 Infestation of *Tunga penetrans*, with secondary infection.

10. Lice

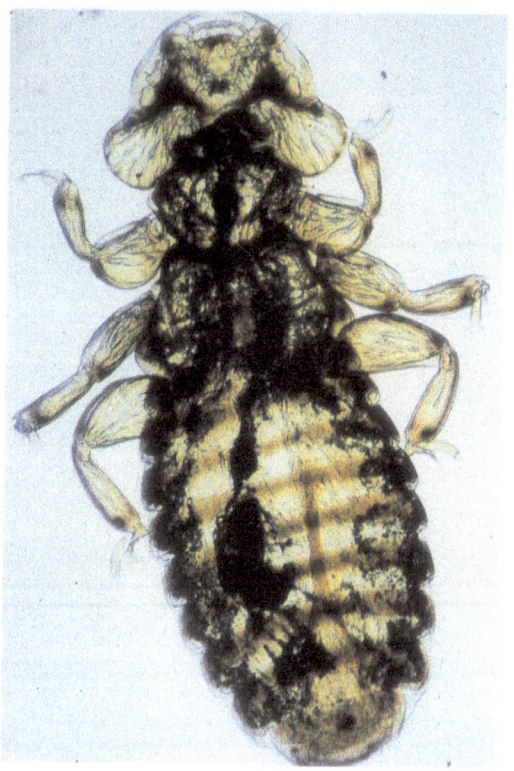

Fig. 10.1   A chewing louse in the order Mallophaga. The head is broader than the abdomen.

Fig. 10.2   A sucking louse in the order Anoplura. The head is narrower than the abdomen.

INTRODUCTION

Lice are permanent ectoparasites of warm-blooded animals. They are remarkably specific in their choice of host, each species of louse being confined to one particular species of animal or bird, away from which they are unable to survive for more than a few hours. Lice are classified according to their feeding habit. Those that chew skin, fur and feathers are in the order Mallophaga (Fig. 10.1), with the head broader than the abdomen, those that suck blood are in the order Anoplura (Fig. 10.2), with the head narrower than the abdomen.

Humans may rarely and only accidentally be attacked by chewing lice. In contrast, three types (species and varieties) of Anoplura will live and feed exclusively on human blood: these are the two varieties of the human louse *Pediculus humanus*: *Pediculus humanus capitis*, the head louse, and *Pediculus humanus corporis*, the body or clothing louse (sometimes classified as two separate species), and the pubic or crab louse *Pthirus Pubis* (sometimes spelt *Phthirus*).

All lice undergo an incomplete metamorphosis in their life cycle and will spend their entire life on the preferred host.

PEDICULUS HUMANUS

Description and life cycle

The two varieties of the human louse are very similar in appearance, the main morphological difference being in the shape of the third segment of the antenna, which is square in the head louse and rectangular in the body louse. There is, however, a considerable difference in the habitat and habits of the two; the head louse is found only on the hair of the head and sucks blood from the scalp, whereas the body louse lives on clothing next to the skin (hence its alternative common name of clothing louse) and feeds on the body.

The human louse (Fig. 10.3) is 2 to 4 mm long, greyish in colour but reddish brown after feeding. It is elongate, wingless and dorsoventrally flattened, with a distinct head (narrower than the thorax), short antennae and a pair of simple eyes. The sucking mouthparts are telescoped inside the head when not in use. The thorax is distinct, with six legs, each terminating characteristically in a large claw. The abdomen is segmented, with a bilobed posterior in the female and a pointed aedeagus (penis) in the male (Fig. 10.4). Head lice are comparatively smaller than body lice.

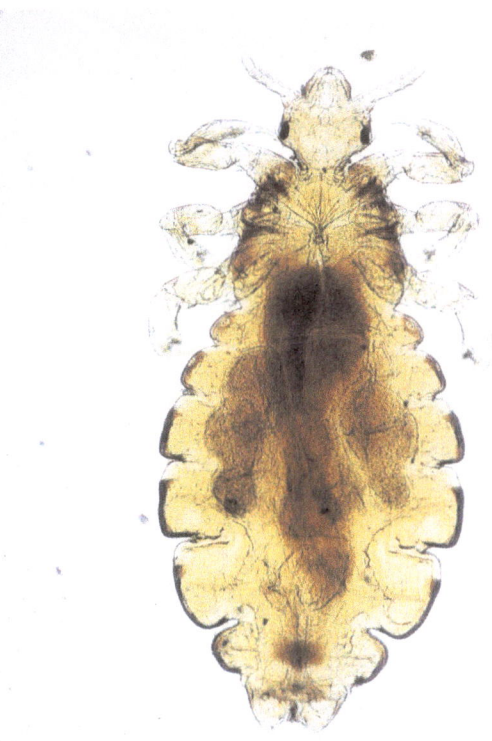

Fig. 10.3   Female head louse.

The gravid female head louse, having mated on the host, will lay about six pinkish coloured eggs per day during her lifetime of 3 to 4 weeks. Each egg is glued firmly to the base of an individual hair (Fig. 10.5) and hatches within approximately 1 week. The pale-coloured empty egg shell, the nit, remains firmly attached to the hair (Fig 10.6), and its distance from the scalp will give a good indication of the length of time of the infestation, since human hair grows at a rate of about 1 cm per month. There are three immature (nymphal) stages, each requiring several blood feeds per day. The adult develops within approximately 10 days (Fig. 10.7).

Head lice are passed from one human host to another in close bodily contact, usually when heads touch. It is unusual for them to be passed via hats,

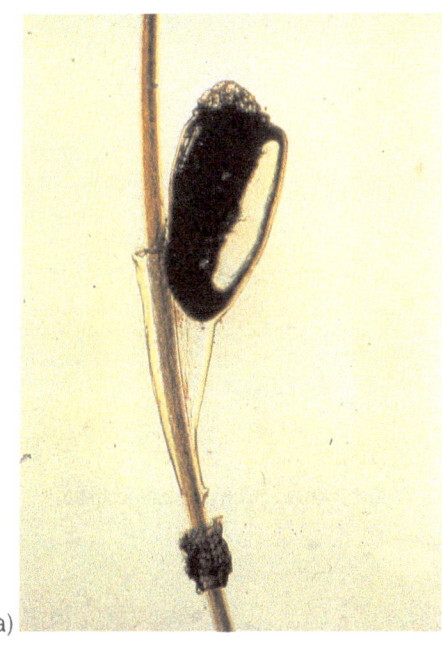

(a)

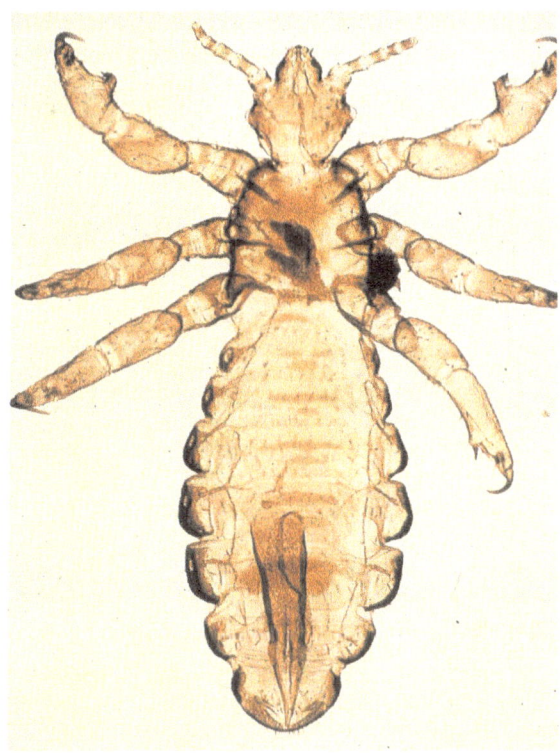

Fig. 10.4   Male head louse.

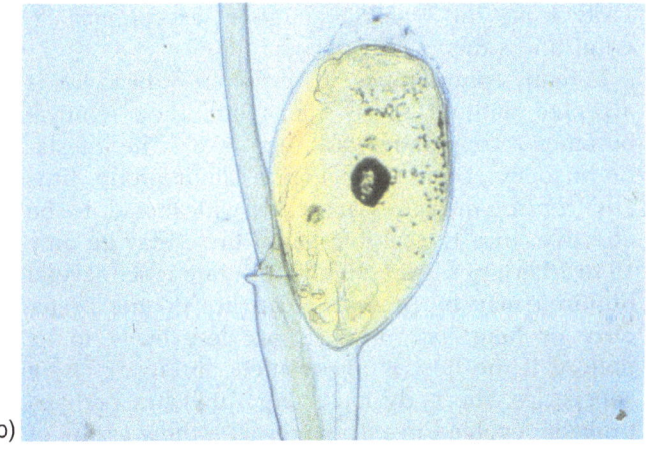

(b)

Fig. 10.5   Head louse eggs. (a) Ovum of the head louse glued to a hair shaft. (b) Developing head louse in egg.

Fig. 10.6 A head of hair heavily infested with head lice. Numerous eggs can be seen.

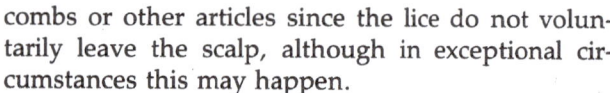

Fig. 10.7 Adult louse and nymphal stages.

Fig. 10.8 Removal of head lice as a cooperative task.

Fig. 10.9 Body louse.

combs or other articles since the lice do not voluntarily leave the scalp, although in exceptional circumstances this may happen.

In many communities, the presence of head lice is accepted without social stigma and their removal becomes a cooperative task (Fig. 10.8). The infestation usually occurs throughout a family group, thus any control measures must be widespread to be effective. In a typical infestation there may be only 10 to 20 lice per head, although in rare cases several hundred may be present. Head lice do not prefer dirty or long hair, but lice are less likely to be noticed if the host is not cosmetically aware. As a subspecies, the body louse (Fig. 10.9) has perhaps probably evolved in association with the wearing of clothes by humans. It lives at the junction between skin and clothing (hence the name clothing louse).

In appearance it is almost identical to the head louse, although slightly larger. Also, the third antennal segment is rectangular rather than square

(Fig. 10.10). The head and body louse share very similar life cycles but their habitats are quite different.

The female body louse will lay eggs on clothing next to skin, particularly on seams and threads, to which they are firmly glued (Fig. 10.11). The young nymph takes a blood feed within a few minutes of hatching and feeds regularly and frequently as it develops through three nymphal stages to the adult. It is unable to survive for more than a few hours if the clothing to which it clings is removed from the body.

Medical significance

The head louse is rarely involved in the transmission of disease, but the body louse is the notorious vector of classical (epidemic) typhus and relapsing fever.

Epidemic typhus is caused by *Rickettsia prowazeki* and is taken up by the louse in a blood meal from an infected human (the only reservoir). The organism multiplies in the gut of the louse and is passed out in the faeces after about 5 days (Fig. 10.12). This is then scratched into a bite puncture or abraded skin of the host. Multiplication of the rickettsiae in the gut cells of the louse is fatal to the vector after about 12 days. Epidemic typhus is a disease of underprivileged situations where the infective louse is able to move from one potential host to another with ease. The disease is more likely to occur in refugee camps, prisons and disaster situations. It is a disease found in many areas of the world (Fig. 10.13).

As with all forms of typhus, clinical aspects include a feverish rash, cough, headache, mental dullness and constipation, often with a macular rash on the trunk. Epidemic louse-borne typhus may be a severe illness complicated by gangrene or venous thrombosis of the limbs. Death may ensue from myocarditis or cerebral involvement, particularly because the victims are already in a debilitated state due to their circumstances. Diagnosis is made serologically, and effective treatment may be achieved with tetracycline or chloramphenicol.

The body louse may be the vector of relapsing fever caused by the spirochaete *Borrelia recurrentis*. The organism is taken up in a blood meal from an infected human host (the only reservoir) and multiplies in the body cavity (haemocoel) of the louse, becoming infective approximately 4 to 6 days later. The organism can be transmitted only when released by crushing the body of the louse, which infested individuals often do. Sometimes lice are crushed between the teeth, which allows the spirochaetes to enter the mucous membranes. The louse will remain infective for a lifetime. Relapsing fever is charac-

Fig. 10.10 Body louse. Note the rectangular third antennal segment.

Fig. 10.11 Body louse. Eggs are laid on clothing that is in contact with skin. The empty egg cases (nits) are very apparent.

Fig. 10.12 Body louse. Rickettsiae causing epidemic typhus are excreted in faeces.

tion of *Borrelia* in stained blood films. Treatment is with penicillin or tetracycline.

PTHIRUS PUBIS

The pubic louse, also known as the crab louse because of its shape, infests the coarser, less-dense hairs of the human body, particularly in the pubic region. It may be found on any hairy part of the body, such as the axillae and legs but very rarely on the hair of the head, which is too dense.

The body of the louse is wide and the claws spaced far apart, thus it requires a suitable space between hairs in order to cling and move satisfactorily (Fig. 10.14). The pubic louse (Fig. 10.15) is

Fig. 10.13 Distribution of louse-borne typhus.

terized by an abrupt onset of high fever with severe headache, painful eye movements, muscle pain, cough, shortness of breath, jaundice and skin haemorrhages. Typically, the illness lasts about 7 days followed by a few days remission and then a relapse, which is less severe. Occasionally, further relapses may occur at intervals. Louse-borne relapsing fever is usually a more severe illness than the form transmitted by soft ticks and is caused by *Borrelia duttoni*. Diagnosis is made by the demonstra-

approximately 2 mm in length, greyish in colour (darker after feeding) and short and broad in shape, somewhat resembling a tiny crab. The head is distinct with obvious antennae and blood-sucking mouthparts, which are telescoped inside it when not in use. The thorax and abdomen are fused and the legs, particularly the middle and hind pair, end in large characteristic claws.

The gravid female louse will glue her eggs to the coarse hairs of the body, often several eggs to a hair

(Fig 10.16). Development to the adult through three nymphal stages takes 10 to 14 days and all stages, including the male and female adult, will feed exclusively on blood.

Medical significance

Sensitization to the bite of the pubic louse may take as long as 30 days to develop (Fig. 10.17), thus it may be a considerable time after initial infestation before their presence is noticed. Heavy and long-term infestations may take the form of dark, pigmented spots and dry skin (Fig. 10.18). The dried blood and louse faeces may be apparent as flecks on bed linen and underclothing. Very occasionally lice may appear in obstetric specimens (Fig. 10.19). Pubic lice may even be found on eyelashes.

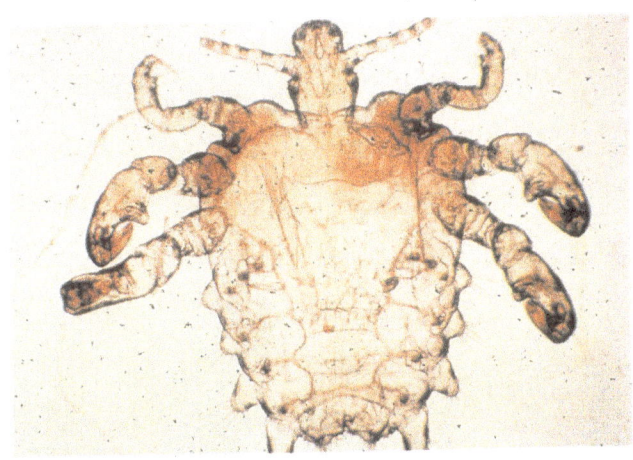

Fig. 10.15 *Phthirus pubis*, the pubic or crab louse.

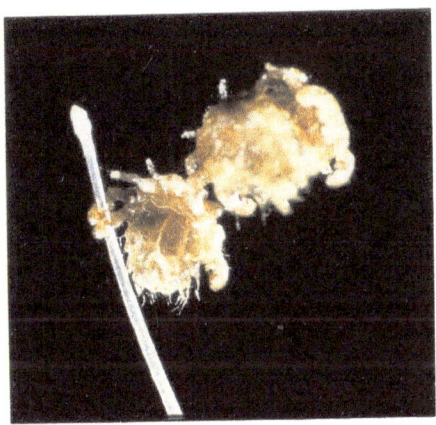

Fig. 10.14 **(a) Crab lice clinging to coarse body hair. (b) Crab louse ova.**

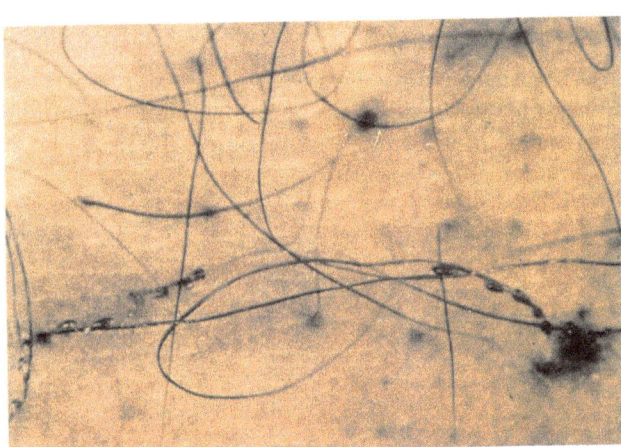

Fig. 10.16 **Several louse eggs glued to the shaft of a pubic hair.**

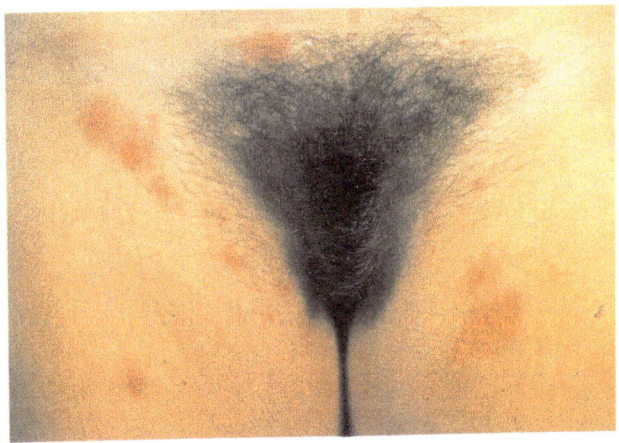

Fig. 10.17 **Sensitization to the pubic louse.** This may take up to 30 days to occur.

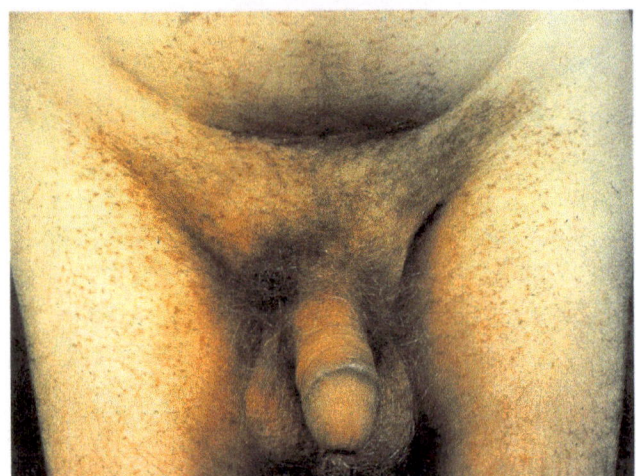

Fig. 10.18   Long-term infestation of the pubic louse showing deeply pigmented spots.

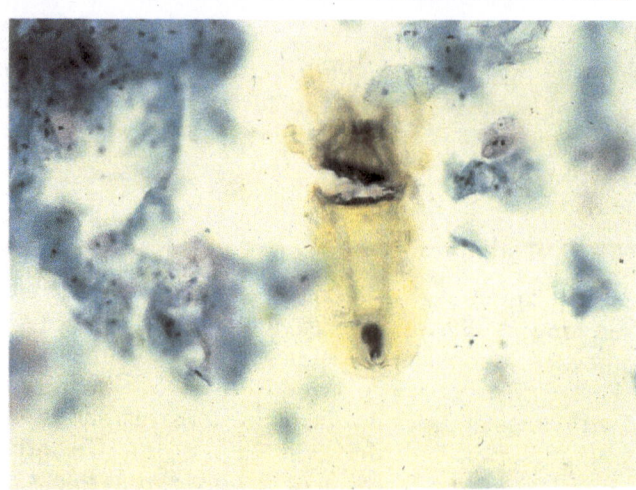

Fig. 10.19   Nymphal crab louse. This was obtained in a cervical smear specimen.

Book lice (order Psocoptera) (Fig. 10.20) are not to be confused with true lice. They do not suck blood but may sometimes be found feeding on moulds growing on discarded skin scales. They are of no public health significance.

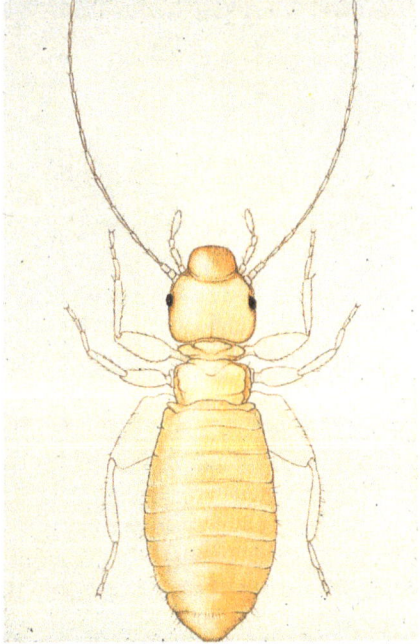

Fig. 10.20   Book lice (order Psocoptera).

11. Bedbugs

INTRODUCTION AND DESCRIPTION

Most bugs, making up the order Hemiptera, are plant feeders. However, two groups, the bedbugs and cone-nosed bugs, have adapted this fluid-feeding ability to sucking blood from birds and mammals, including humans.

Within the family Cimicidae are the blood-sucking bugs associated with nests of birds such as swallows, as well as those of bats. Also within this family are the bedbugs that feed on humans, *Cimex lectularis* (Fig. 11.1), which is found throughout the temperate regions, and *C. hemipterus* (= *C. rotundatus*), which is found mainly in the tropics and subtropics. Their appearance is very similar, the most obvious difference being the presence of 'cusps' (protruding shoulder pads) on the prothorax of *C. lectularius*.

BREEDING SITES

Bedbugs are temporary ectoparasites of humans, feeding mainly nocturnally on the sleeping host. When not feeding they will hide in cracks and crevices, behind furniture, fixtures and other harbourages in human habitation. Their presence can sometimes be sensed by a sweet sickly smell but much more apparent is the spotting of surfaces around the harbourages caused by their faeces (Fig. 11.2).

(a)

(b)

Fig. 11.2 Bedbug harbourages. There is typical faecal spotting of the wall panelling(a), and on the surface of a bedhead (b).

DESCRIPTION AND LIFE CYCLE

Cimex (Fig. 11.3) is an oval flattened insect of 4 to 5 mm in length, and pale brown in colour although it appears darker after a blood meal. The head is short and broad, with a pair of prominent antennae and a pair of small, dark, compound eyes. The long, slender mouthparts are slung beneath the head and thorax when not in use but are rotated downwards when feeding (Fig. 11.4). The bedbug is secondarily wingless, having lost the ability to fly during its evolution as a blood-sucking ectoparasite. Three pairs of well-developed legs enable the bedbug to crawl rapidly. The abdomen is large and conspicuously segmented, with a rounded posterior in the female (Fig. 11.5). The male has a more pointed posterior with a curved protruding penis or aedeagus at the tip (Fig. 11.6).

The bedbug is closely associated with humans and the domestic environment throughout its life cycle, which is one of incomplete metamorphosis (Fig.

Fig. 11.1 Female *Cimex lectularius*, unfed. Note the shoulder cusps typical of this species.

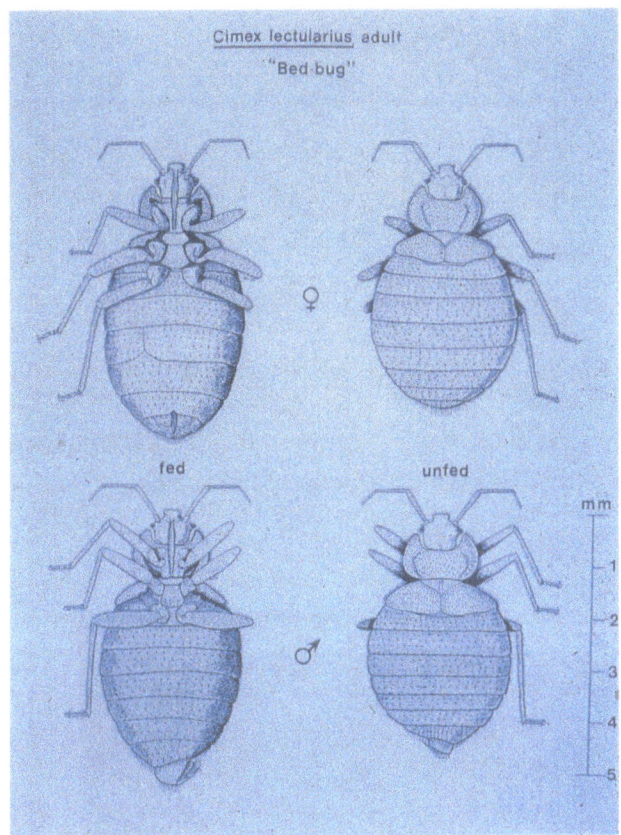

Fig. 11.3 Male and female bedbugs.

Fig. 11.4 Bedbug taking a blood meal. Note the position of the mouthparts.

Fig. 11.5 Female bedbug.

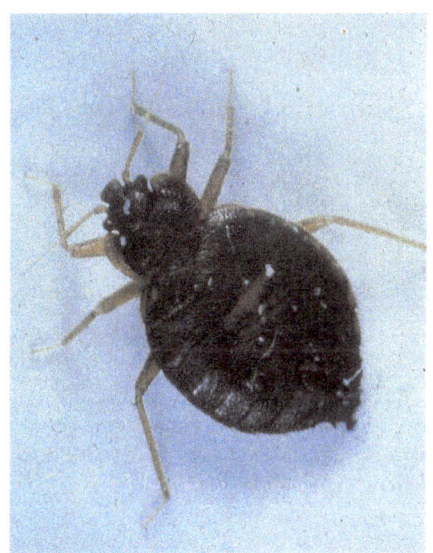

Fig. 11.6 Male bedbug with protruding aedeagus.

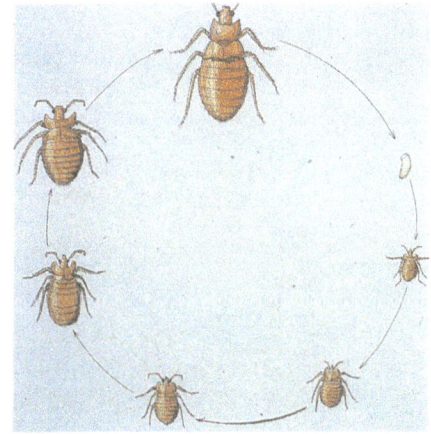

Fig. 11.7 Life cycle of the bedbug.

11.7). After mating, the female will lay two to three eggs per day throughout her life, which may last several months, thus several hundred eggs may be laid during a lifetime. Eggs will be laid in harbourages (Fig. 11.8); they are about 1 mm long, yellowish-white and vase-shaped with a lid or operculum at one end (Fig. 11.9). They hatch within 10 days at room temperature (20°C), or sooner at higher temperatures, but become non-viable below 14°C. There are five immature (nymphal) stages, each taking at least one nocturnal blood meal, and developing to the adult in about 6 weeks under optimal conditions. This takes considerably longer at lower temperatures or in the absence of blood. Adults, both male and female, will feed regularly and frequently, taking several minutes to complete a blood meal, and while human blood is preferred they will also feed on rodents, rabbits, bats and birds if necessary.

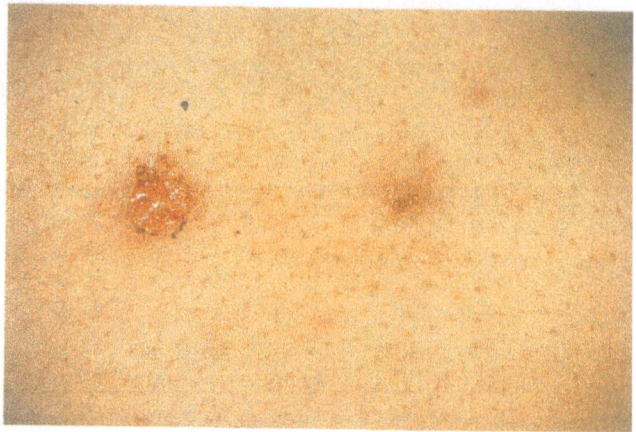

Fig. 11.10   Typical excoriated bedbug bite.

Fig. 11.8   Bedbug infestation. A prison cell had become infested, the bedbugs laying eggs in the mortar cracks on the wall. Nymphs and cast skins can also be seen.

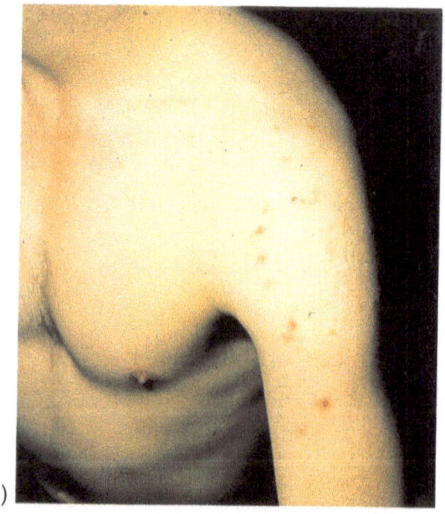

(a)

Fig. 11.9   Bedbugs and eggs. These were located on a screw hole in a bed frame.

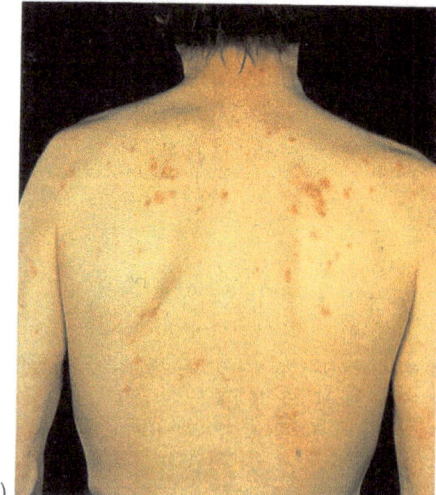

(b)

Fig. 11.11   Typical appearance of bedbug bites on parts of the body exposed at night: (a) on the shoulders; (b) on the back.

MEDICAL SIGNIFICANCE

Although all stages in the life cycle of bedbugs rely exclusively on blood for nutrition, they are not natural vectors of any human disease. However, their bite may cause considerable irritation resulting in excoriation of the skin (Fig. 11.10) and swelling in sensitized victims. Their nocturnal blood-sucking habits may result in significant loss of sleep and diminished health and morale. Being wingless, they rely on being able to crawl or drop onto their host, thus bites will commonly and characteristically occur on parts of the body uncovered at night, especially the arms and the shoulders (Fig. 11.11). Heavy infestations are usually associated with substandard domestic conditions and poor hygiene, and only under very extreme conditions will the bedbugs be found harbouring on the victim (Fig. 11.12). Minor infestations with the presence of a few insects may be the result of articles transported from an infested area.

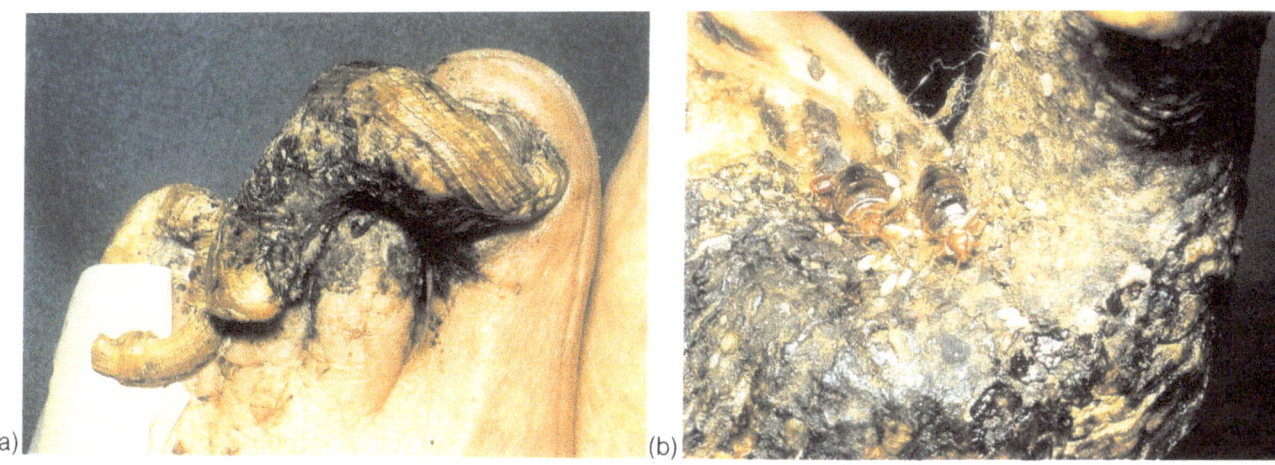

(a) (b)

Fig. 11.12 Extreme bedbug infestation. The victim was semi-destitute, living in a room which contained thousands of bedbugs. The feet were in poor condition with uncut toenails (a), and bedbugs were harbouring between the toes (b).

12. Cone-nose bugs

INTRODUCTION AND DESCRIPTION

Cone-nose bugs form the subfamily Triatominae (triatome bugs) within the family Reduviidae (reduvid bugs), and are found almost exclusively in the New World. They are all blood-sucking, temporary ectoparasites of warm-blooded animals. Species in three genera, particularly *Rhodnius prolixus*, *Triatoma infestans* and *Panstrongylus megistus*, will feed habitually on humans and may transmit *Trypanosoma cruzi*, the causative organism of American trypanosomiasis or Chagas' disease.

Cone-nose bugs are of medium to large size, 10 to 25 mm in length and brown in colour, often with red, yellow, black or white markings (Fig. 12.1). The shape of the head and prothorax gives the bug its common name of 'cone-nose'. The elongate head has a small dark compound eye on either side, and the long thin proboscis is hinged beneath it when not in use, stretching below the thorax. The antennae are long and have four segments. Their position in relation to the eyes and the tip of the head is characteristic of the genus (Fig. 12.2). The prothorax is prominent. Three pairs of strong legs and two pairs of wings in the adult arise from the thorax. The front pair of wings are heavily veined and pigmented, and are folded in a scissor-like fashion over the membranous hind pair when at rest. Adults are able to fly efficiently. The abdomen is large and flat when unfed but swells considerably when full of blood (Fig. 12.3).

(a)

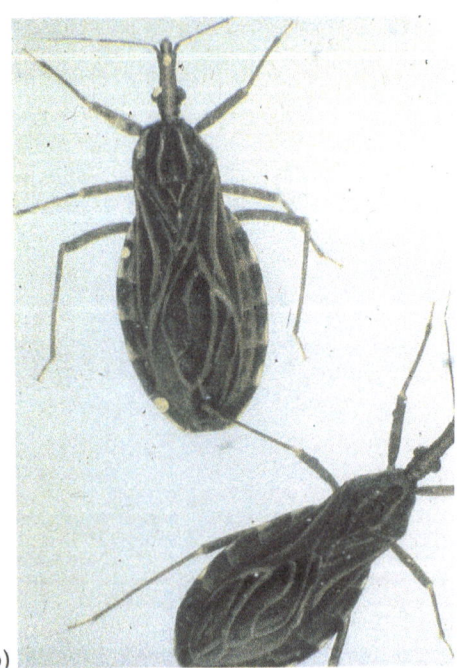

(b)

Fig. 12.1 Cone-nose bugs. (a) *Triatoma infestans.* (b) *Rhodnius prolixus.*

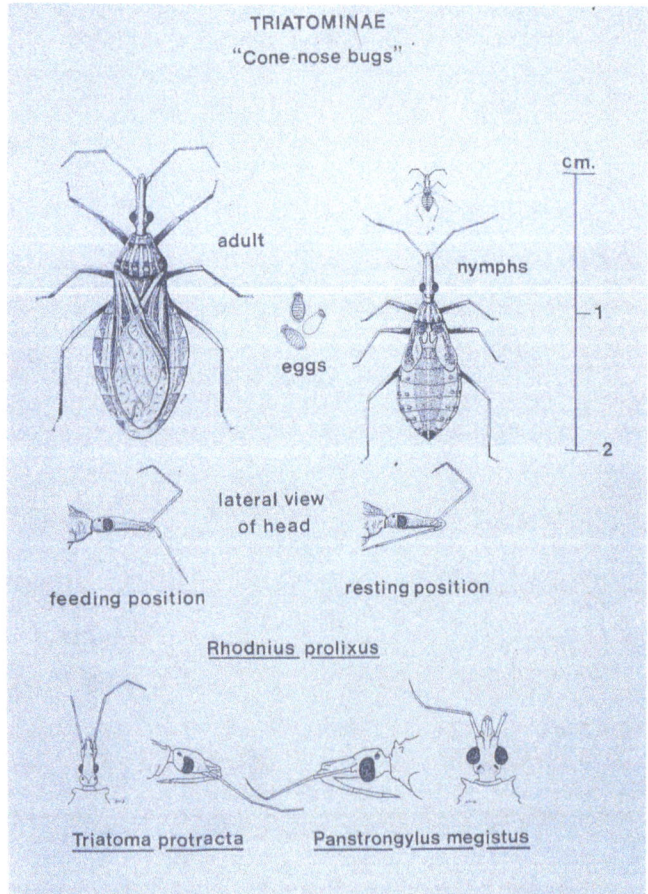

Fig. 12.2 Cone-nose bugs showing differentiation of genera.

Fig. 12.3 Blood-fed cone-nose bugs. The faeces have been produced while the bug feeds.

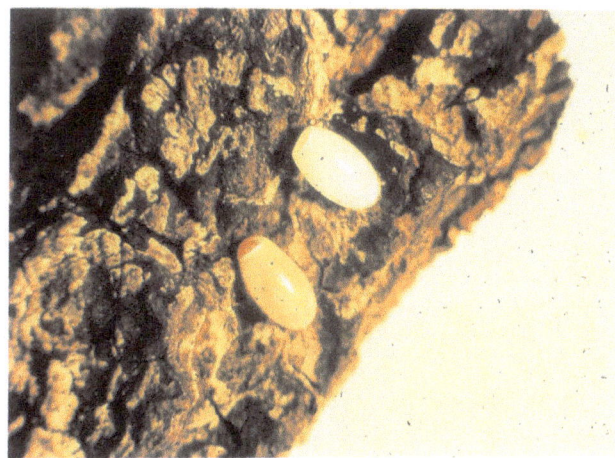

Fig. 12.5 Triatome eggs.

LIFE CYCLE

Cone-nose bugs are associated particularly with rodents and other small rural burrowing animals. Their life cycle, which is one of incomplete metamorphosis, may be completed in the host's nest or burrow or in caves inhabited by the host. Infestations may be established in human habitation by the flying gravid female, particularly where the dwelling provides adequate harbourages in wooden walls and atap roofing (Fig. 12.4). About 2 weeks after mating and taking the first blood meal, the gravid female will lay a total of about 200 eggs in batches of ten to 20, in cracks and other harbourages. The egg is flask-shaped, about 2 mm in length with a well-marked lid or operculum (Fig. 12.5). Most species that infest human habitation lay pinkish-coloured eggs. Once the egg has hatched, about 10 days after being laid, the empty egg case is pale in colour (Fig. 12.6). The nymph that emerges (Fig. 12.7) is wingless and will shelter for a few days before taking its first blood meal (Fig. 12.8), which may be many times its own weight. There are five nymphal stages, each resembling a mature adult except for the absence of wings. The immature cycle may cover several months and the adults may live for at least 2 years.

MEDICAL SIGNIFICANCE

The bite of the cone-nose bug may be painless, hence the common name 'kissing bug', but it may cause urticaria in a sensitized host. The bug is the

Fig. 12.4 Infestation of wooden walls and atap roofs in a typical dwelling. The bugs will emerge at night and feed on the inhabitants.

Fig. 12.6 Egg and colourless empty egg case.

Fig. 12.7   Nymph of *Triatoma infestans*.

Fig. 12.9   Distribution of Chagas' disease.

Fig. 12.8   Nymph feeding.

vector of *Trypanosoma cruzi*, the causative organism of American trypanosomiasis or Chagas' disease in Central and South America (Fig. 12.9). The organism is taken up in a blood meal from an infected host reservoir, which may include rodent, armadillo, cat or dog, and undergoes development in the gut of the bug. It is passed to humans some 6 to 15 days later when the bug defaecates while feeding. The organism is scratched into the bite puncture or passes through the mucous membrane of the lips, nose, or eyelid on which the bug feeds.

American trypanosomiasis is common in northern South and Central America, especially in Brazil. The site of inoculation of the parasite by the bite of the bug, often occurring on the head of a child, may form a blind boil (chagoma), with accompanying allergic oedema. Dissemination of the parasite may lead to fever and enlargement of the lymph nodes, liver and spleen, and may cause death from myocarditis or meningoencephalitis. The patient may recover and remain well for years, only to suffer damage to heart muscle (leading to heart failure or sudden death) or to the nerve plexuses in the gastrointestinal tract, leading to gross dilatation of the oesophagus or the colon.

Diagnosis in the early stages of the disease can be made by the discovery of trypanosomes in stained blood films, but in later stages diagnosis can only be made by serological methods. Laboratory-bred cone-nose bugs may be used in the technique of xeno-diagnosis. Trypanosome-free bugs are fed on a suspected patient and the organisms, if present, will become concentrated in the hindgut of the bug, the contents of which are examined under a microscope after about 7 days. Treatment (with nifurtimox and primaquine) is only effective in the early illness.

13. Cockroaches

INTRODUCTION AND DESCRIPTION

The order Dictyoptera comprises cockroaches and mantids. A few species have become domestic pests worldwide and, because of their close association with humans, food and infected material, have become incriminated in the mechanical transmission of disease. There are some 4000 species of cockroach distributed throughout most parts of the world, although most are of little or no significance to humans. Many species are diurnal in habit and a large proportion live in tropical rain forests. They are closely related to the mantids (Fig, 13.1) and were, until recently, classified in the order Orthoptera with the grasshoppers and crickets. They are perhaps one of the most successful of all insect groups and have changed little in appearance in 250 million years. Approximately 50 species have acquired the habit of domestication to a greater or lesser extent, and some have followed humans to most parts of the globe, becoming widespread and significant domestic pests.

Cockroach species vary in size from 5 mm to 9 cm, and those species that have become domestic pests range from 10 to 50 mm. Most of these are tropical in origin and have adapted to living in a pseudotropical environment provided by humans in temperate regions: in kitchens, restaurants and food-processing premises. The most important pest species are *Blattella germanica*, the German cockroach (Fig. 13.2), *Blatta orientalis*, the oriental cockroach (Fig. 13.3), and *Periplaneta americana* (Fig. 13.4), the American cockroach.

The adult German cockroach is 12 to 15 mm in length and is pale to mid-brown, with two dark longitudinal bands on the prothorax. Both male and female adults have wings although they do not fly in temperate regions. The female carries the eggs enclosed in an ootheca until just before they are laid.

The adult oriental cockroach is 20 to 25 mm long, broad in shape and dark brown to black in colour. The wings are greatly reduced in the female and vestigial in the male.

The adult American cockroach is 30 to 45 mm in

Fig. 13.1   Praying mantis.

Fig. 13.3   *Blatta orientalis.*

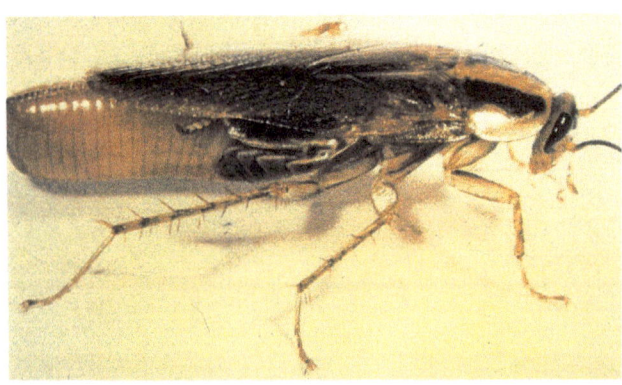

Fig. 13.2   *Blattella germanica.*

Fig. 13.4   *Periplaneta americana*, a typical cockroach.

length, reddish brown in colour with prominent yellow patches on either side of the pronotum (the dorsal prothorax). Both male and female have large wings overlapping the abdomen.

The body of a typical pest species of cockroach (see Fig. 13.4) is flattened dorsoventrally and clearly divided into head, thorax and abdomen. The front dorsal part of the thorax is well developed and shelters the head from above. The head bears a pair of long filamentous antennae, paired compound eyes and chewing mouthparts that are adapted for omnivorous feeding and are slung ventrally. The thorax holds three pairs of strong legs and, in most species, two pairs of wings folded in a scissor-like fashion over the abdomen. The abdomen is well-segmented and bears paired appendages (cerci) at the hind end.

LIFE CYCLE

The life cycle is one of incomplete metamorphosis. Some days after mating the female will lay eggs in an ootheca (a feature of all dictyopterans), which may contain 12 to 50 eggs depending on the species. The ootheca is very resistant to desiccation, and is often laid in a protected harbourage such as a hollow space behind panelling (Fig. 13.5). The first-stage nymph, which hatches from each egg some 2 to 6 weeks later, is a wingless miniature of the adult with a soft and pale exoskeleton (Fig. 13.6). On exposure to air the exoskeleton hardens and darkens. Depending on the species there may be five to 12 nymphal stages before the final moult to the typically winged adult (Fig. 13.7). All stages of cockroach will feed on a wide range of substances, particularly refuse, detritus and human food. They will be found in drainage systems where there is ready access to potentially infected material in the form of faeces (Fig. 13.8), and from where the cockroach may be able to reach domestic situations (Fig. 13.9). The method of feeding involves regurgitation of gastric fluid and external maceration before swallowing the pulped material. In doing this, a residue may be left from a previous meal of contaminated material. The external body and legs are clothed with stout bristles and hairs to which particles of detritus may adhere. In a domestic situation, cockroaches will feed at night on refuse and debris found on floor surfaces,

While feeding and moving over cutlery and plates, defaecation is indiscriminate (Fig. 13.10). During feeding, the cockroach also preens itself to remove

Fig. 13.5 **Oothecae of the oriental cockroach.** These were removed from a hollow door frame.

Fig. 13.6 **Hatching ootheca.**

Fig. 13.7 **Cockroach nymph moulting into the adult form.**

a)

(a)

b)

(b)

Fig. 13.8 Habitats of the cockroach. (a) In tropical regions, *Periplaneta americana* are often found in sewers. (b) This pit latrine in Botswana harboured *Periplaneta americana*. The area also supported fly larvae and pupae.

(c)

Fig. 13.10 Cockroach infestations. (a) In kitchens, cockroaches may feed on debris in corners of a floor surface. (b) Food left unprotected at night in areas such as hospital and hotel kitchens will encourage infestation. (c) Cockroaches crawling over cutlery and crockery may cause contamination.

Fig. 13.9 Infestation of domestic habitats. Cockroaches may gain access to a kitchen from the sewer via a damaged water trap. Drainage in the kitchen provides water which is most important for survival.

debris clinging to the legs and body. The presence of dark unsightly smears on wall surfaces and around harbourages is diagnostic of cockroach infestation (Fig. 13.11). In this manner, contaminated food, utensils and surfaces will become a potential source of infection to humans. Table 13.1 shows a list of organisms pathogenic to humans, which have been isolated from cockroaches collected in hospitals and domestic situations in which patients were suffering from disease. Table 13.2 shows that cockroaches will take up, and retain in a viable state, a wide range of organisms in their environment. Figure 13.12 shows a MacConkey medium plate on which a cockroach infected with *Escherichia coli* has defaecated and spread faeces for 30 seconds. The incubated medium shows its path clearly.

Many food materials provide good culture media for enteric organisms, particularly in the warm atmosphere of a kitchen.

Fig. 13.11   Evidence of cockroach infestations.

Fig. 13.12   Incubated MacConkey medium plate on which a cockroach has spread its faeces containing *Escherichia coli*.

Table 13.1 Pathogenic organisms found naturally infecting cockroaches	
Organism	Disease
BACTERIA	
Pseudomonas aeruginosa	Infections of: Urinary tract Upper resp. tract Wounds, burns
Staphylococcus aureus	Boils, abscesses
Streptococcus faecalis	Faecal contamination
Escherichia coli	Urogenital/intestinal infections
Salmonella spp incl. S. typhi & S. typhimurium	Typhoid, other enteric fevers
	Gastroenteritis
Shigella spp	Dysentery, diarrhoea
Klebsiella pneumoniae	Pneumonia/URTI
Serratia marcescens	Upper resp. tract. inf.
Proteus vulgaris	Gastroenteric tract inf.
Yersinia pestis	Plague
FUNGI	
Aspergillus fumigatus	Aspergillosis
PROTOZOA	
Entamoeba histolytica	Dysentery
HELMINTHS (worms)	
Enterobius vermicularis	Pin/threadworm
Trichuris trichiura	Whipworm
Ascaris lumbricoides	Roundworms
Ancylostoma duodenale Necator americanus	Hookworms
VIRUSES	
Hepatitis Poliomyelitis	'Jaundice'

MEDICAL SIGNIFICANCE

The characteristics of the organisms detailed in Table 13.1 make it almost impossible to confirm that a cockroach is the prime cause of any disease outbreak. By their very nature these diseases can be transmitted in a variety of ways and it is unlikely that the cockroach is the specific carrier. However, circumstantial evidence and scientific commonsense indicate that, where cockroaches have access to potentially infected material and to human food and food preparation areas, their role in the transmission of disease should not be ignored.

Cockroach debris, dried cast skins and faeces may cause an allergic response in susceptible patients.

This may take the form of a dermatitis; it may also present as a respiratory condition or with watering eyes and nose.

Table 13.2 Enterobacteriaceae found in Periplaneta Americana from insectary

Organisms	Found in cockroach	Persistent for up to (days)	Found in cockroaches dead for 10 days	Found in environment
Citrobacter diversus	√	10	√	√
Citrobacter freundii	√	10	√	√
Enterobacter aerogenes	√	–	–	√
Enterobacter aglomerans	√	7	–	√
Enterobacter cloacae	√	–	–	√
Enterobacter hafniae	√	7	–	√
Escherichia blattae	√	9	–	√
Escherichia coli	√	2	–	√
Klebsiella pneumoniae	√	8	√	√
Proteus mirabilis	√	6	–	√
Proteus morganii	√	8	√	√
Proteus rettgeri	√	6	√	√
Proteus vulgaris	√	10	√	√
Serratia marcescens	√	–	√	√

14. The Arachnids

INTRODUCTION AND DESCRIPTION

In addition to the insects, the phylum Arthropoda contains several other classes, of which the most medically significant is the Arachnida, which includes spiders, scorpions, ticks and mites. All arachnids have a body that is divided into a fore- and hind-part, which may be fused into a sac-like structure. There is no separate head and no antennae. Mouthparts consist of paired chelicerae and pedipalps, and sometimes a central hypostome, all of which may be varied in appearance depending on feeding habits. The adult arachnid has four pairs of legs attached to the front part of the body, but wings are never present. Some arachnids possess venom glands incorporated in the mouthparts (spiders, Fig. 14.1), or at the hind end of the body attached to a needle-like sting (scorpions, Fig. 14.2). Some mites (Fig. 14.3) will cause an allergic skin or respiratory reaction (e.g. house dust mite and storage mite), others will burrow into the tissue (e.g. scabies mite) or may transmit disease (e.g. scrub typhus mite). Ticks (Fig. 14.4) and a few mites will feed exclusively on blood and may act as vectors of a range of diseases.

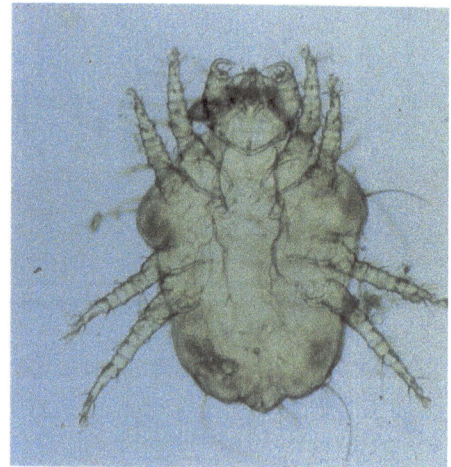

Fig. 14.3   Mite.

Fig. 14.1   Spider.

Fig. 14.4   Tick.

TICKS

The order Acarina consists of ticks and mites. These can be differentiated by the larger size of the ticks and the presence of a spiracle on either side of the body near the hind pair of legs. All ticks feed on blood from amphibians, reptiles, birds and mammals, and have a strongly developed hypostome (Fig. 14.5), which is used to anchor the other paired

Fig. 14.2   Scorpion.

Fig. 14.5 Hypostome of a hard tick.

mouthparts to the host. There are two distinct groups (families) of tick: the hard ticks (Ixodidae), of which there are some 650 species in 11 genera, which spend most of their life attached to their host; and the soft ticks (Argasidae) of which there are 150 species in four genera, which live in the habitat of the host, only parasitizing during a blood feed.

Hard ticks

Adult hard ticks are oval or pear-shaped with a sac-like body, and are dorsoventrally flattened. The length varies from 2 to 10 mm when unfed, although the fully gorged female may be at least twice this size. The common name derives from the chitinized shield that covers the dorsal surface of the male and is often characteristically patterned, but only extends over the anterior dorsal surface of the female (Fig. 14.6). This allows the female to accommodate the large blood meal. The mouthparts of both male and female are clearly visible from above. The eight legs of the adult are large and attached to the front part of the body.

Life cycle

Almost the whole life cycle of the hard tick is spent on the host. Those that parasitize humans will be more commonly found on sheep, cattle, deer, rodents and dogs, and include species of *Ixodes*, *Dermacentor*, *Amblyomma* and *Haemaphysalis*. Having mated on the host, the gravid female drops to the ground, lays a massive batch of several thousand eggs (Fig. 14.7) and then dies. The eggs will usually hatch within a few days but may overwinter depending on temperature and humidity. From the egg will emerge a six-legged larva (Fig. 14.8), which will climb onto vegetation and wait for a host to pass. It will typically wave its front legs in the air to detect the host and then cling to its fur as it passes, crawling immediately to a soft area such as the axilla or ear (Fig. 14.9) where it takes a blood meal. The larva then moults to an eight-legged nymphal stage (Fig. 14.10); some species do this on the host, others drop to the ground and will find a new host after moulting. The nymph takes a blood meal, which will last for several days. It will eventually moult into an

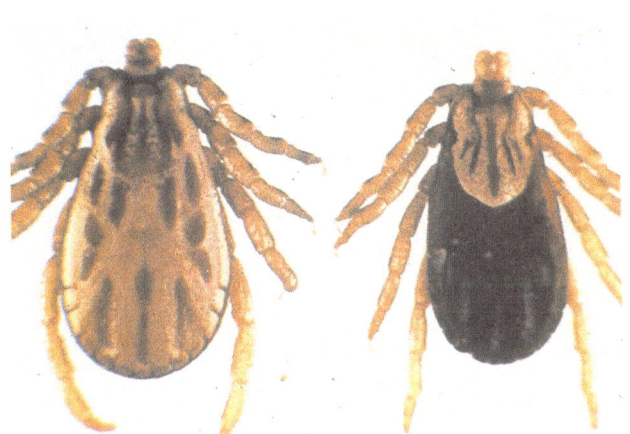

Fig. 14.6 Male and female hard ticks.

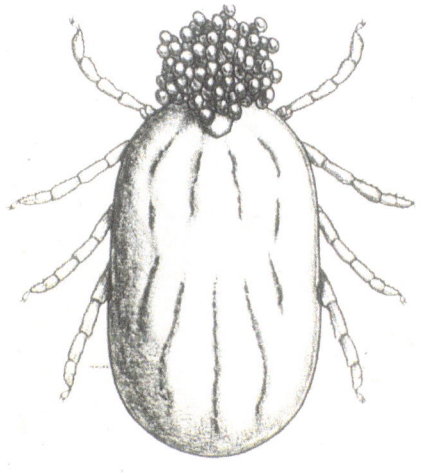

Fig. 14.7 Egg batch of hard tick.

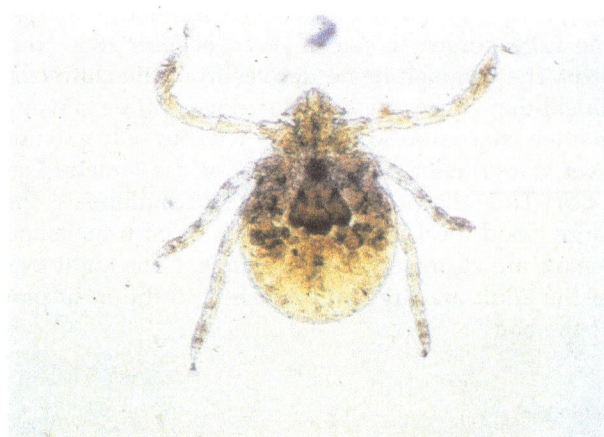

Fig. 14.8   Hard tick larva.

Fig. 14.9   Hard tick larvae on rabbit's ear.

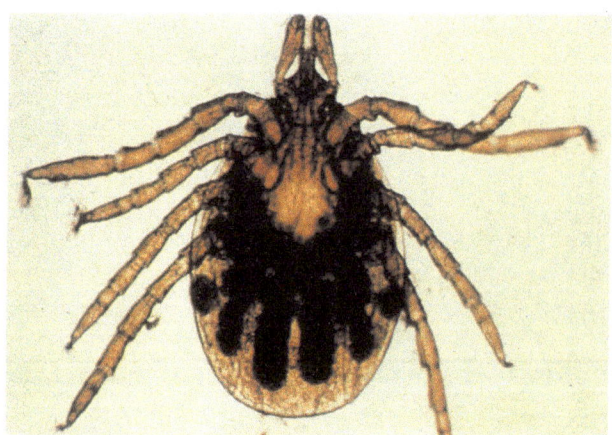

Fig. 14.10   Ventral surface of hard tick showing genital orifice.

adult, on or off the host depending on the species. The blood feed of the adult, particularly that of the female, make take up to 3 weeks to complete, and the body weight may increase by a factor of 200. The body may be so swollen with blood that the dorsal shield may be difficult to see (Fig. 14.11). While the female is feeding, the male tick, having completed a blood meal, will move over the host searching for a female of the same species, and mating will occur on the host. The genital opening of both sexes is close to the base of the mouthparts on the ventral surface (Fig. 14.12), thus transfer of sperm is accomplished by the male crawling underneath the female while she continues to feed.

Those species of tick that leave the host at each stage to moult (larva to nymph, nymph to adult) are known as 'three-host' ticks since they parasitize three different individual animals as larva, nymph and adult. These include the more important disease vectors such as *Ixodes ricinus* and *Dermacentor andersoni*, whose habits increase their capacity as vectors. Some *Hyalomma* and *Rhipicephalus* species remain on

(a)

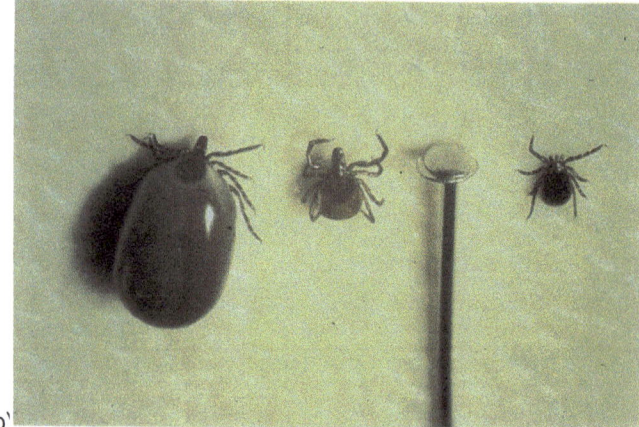

(b)

Fig. 14.11   Fully fed female tick. (a) With larva. (b) With nymphs.

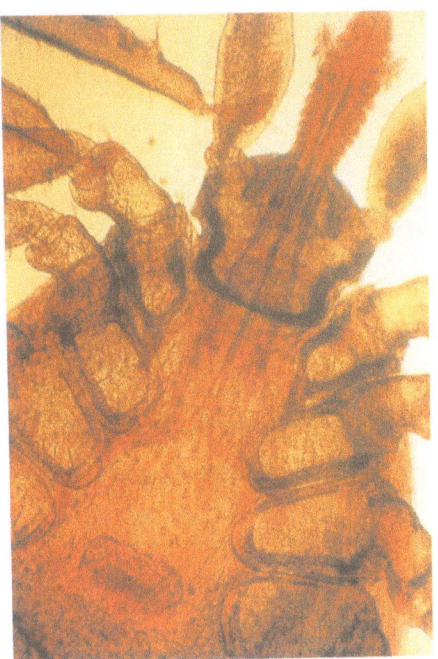

Fig. 14.12   The genital opening of the hard tick. In both sexes the genital opening is close to the base of the mouth parts on the ventral surface.

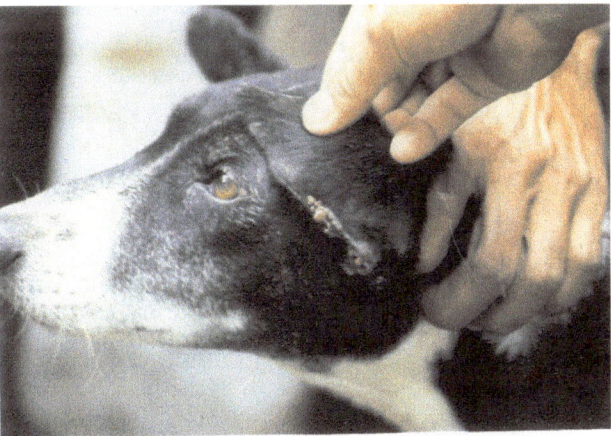

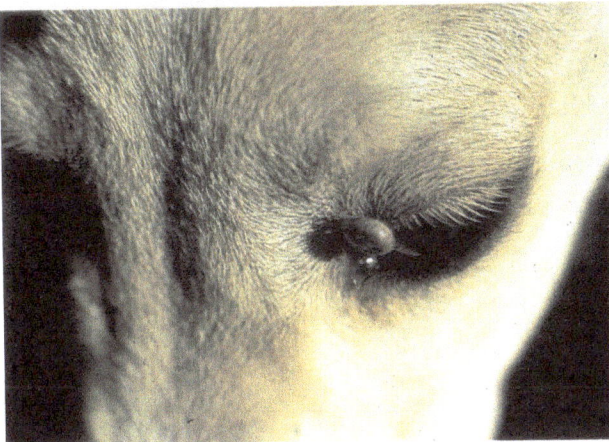

Fig. 14.14   Hard tick feeding on domestic animal. (a) Feeding on the ear. (b) Feeding on the eyelid.

the host when moulting from larva to nymph and are thus known as 'two-host' ticks, and some *Boophilus* species only leave the host at the end of their adult life and are thus known as 'one-host' ticks.

Medical significance

Species of hard ticks at all stages will feed on humans (Fig.14.13), causing irritation and sometimes alarm as they swell considerably in size. The

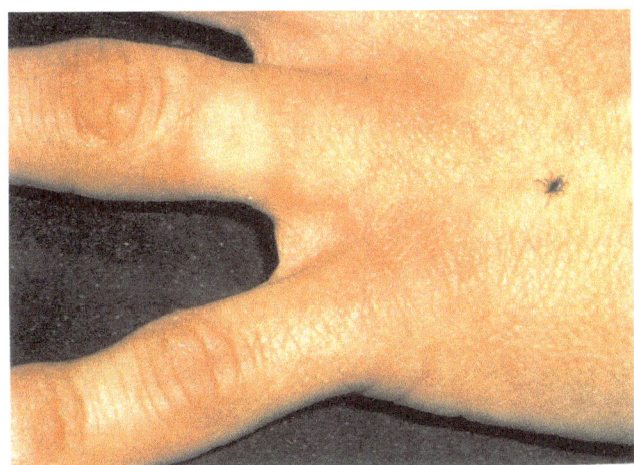

Fig. 14.13   Hard tick feeding on human.

ticks may be brought into human habitation by domestic animals (Fig. 14.14). They may leave the host to moult and may parasitize humans at the next stage of the life cycle. If the tick is removed clumsily, the mouthparts, particularly the hypostome, will be left in the bite puncture and may give rise to secondary infection. The feeding ticks should be removed by holding the body with forceps (Fig. 14.15), gently pushing the body into the tissue to release the barbed hypostome, twisting the body and pulling away. Alternatively, the tick may remove itself if the air supply is cut off, by covering it with liquid paraffin or by irritating it with insect repellent or a lighted cigarette.

Ticks are often acquired when walking through infested vegetation that is inhabited by the normal host, and large numbers, particularly of larvae, may be picked up along habitual animal runs. A practical method for collecting ticks involves dragging a rough sheet or blanket through an infested area and picking off the ticks that cling to it (Fig. 14.16).

Hard ticks may act as vectors of a wide variety of

Fig. 14.15   Tick removal with forceps.

Fig. 14.16   Blanket-dragging for ticks.

disease organisms including several encephalitic and haemorrhagic arboviruses (Fig. 14.17), rickettsial diseases such as tick typhus, boutonneuse fever and Q-fever (Fig. 14.18), as well as Lyme disease. While feeding, the female hard tick may also cause a form of ascending motor paralysis.

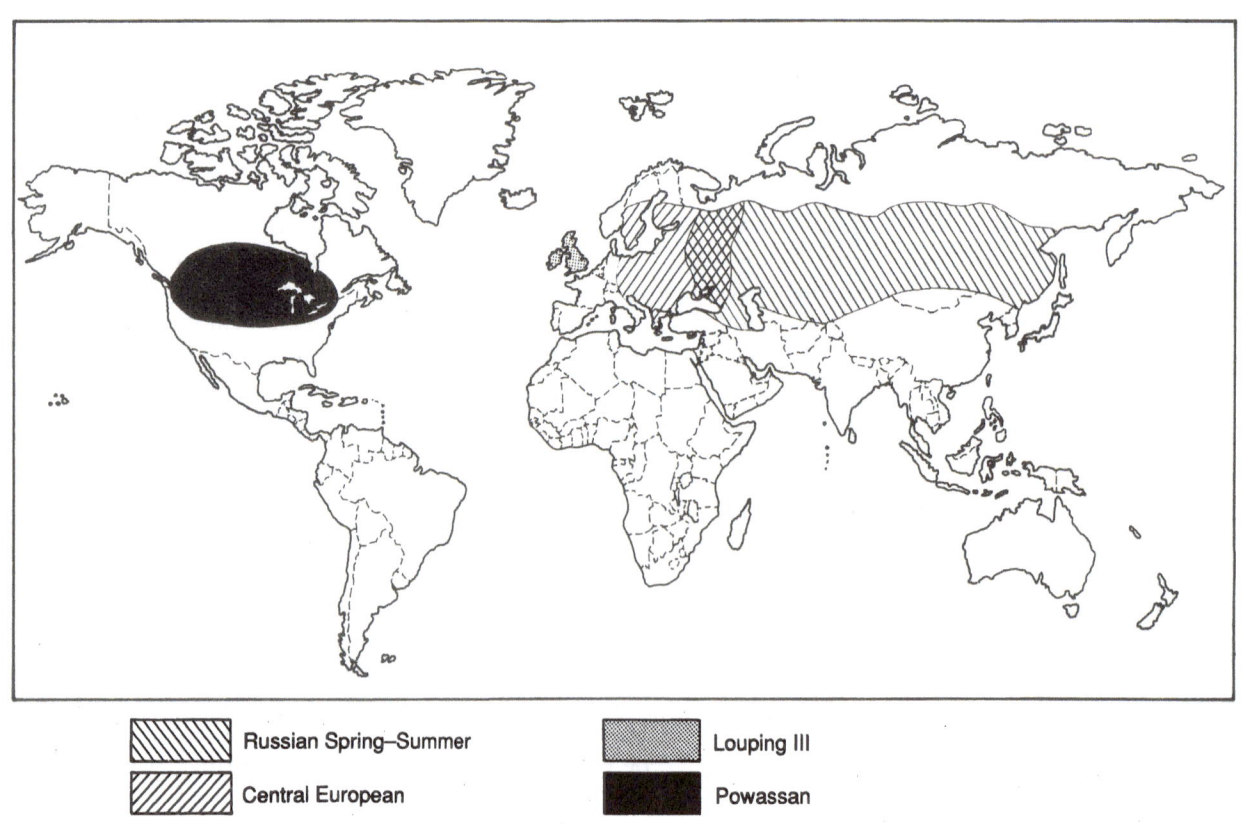

Russian Spring–Summer

Central European

Louping III

Powassan

Fig. 14.17a   Distribution of encephalitic tick-borne viruses.

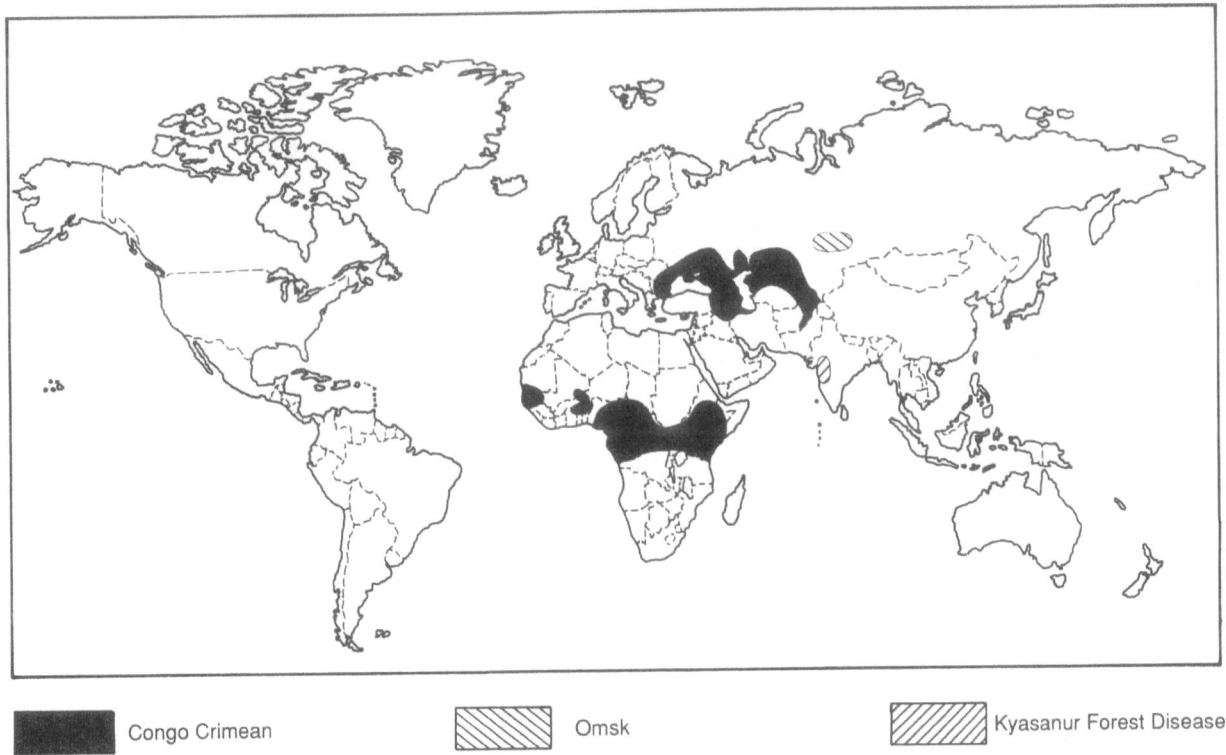

| Congo Crimean | Omsk | Kyasanur Forest Disease |

Fig. 14.17b   Distribution of haemorrhagic tick-borne viruses.

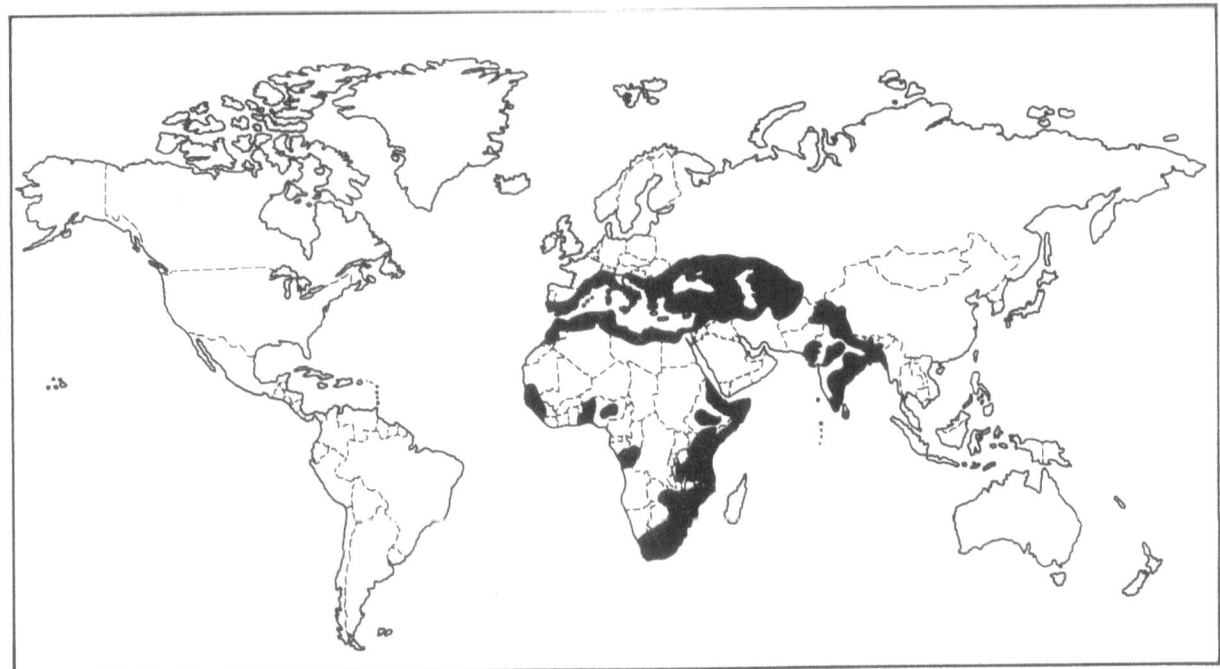

Fig. 14.18   Distribution of tick-borne rickettsial diseases.

Tick-borne diseases

Arboviruses

Viruses carried by ticks are exemplified by those which cause encephalitis in Russia and Siberia and in South East Europe, and the haemorrhagic fever syndrome named Congo-Crimea haemorrhagic fever which is transmitted by ticks from ruminant animals to man in Central Africa, Arabia, Crimea and Ukraine.

Tick typhus

All forms of typhus produce a feverish illness accompanied by cough, headache, mental dullness and constipation. A macular rash may occur on the trunk. A small black scab apparent at the site of the bite is seen in tick-borne typhus in Brazil, the Mediterranean, Africa and India. However, a haemorrhagic rash, known as Rocky Mountain spotted fever, is the hallmark of tick typhus in North America. Diagnosis of rickettsial diseases is carried out serologically, and treatment is effective with tetracylines or chloramphenicol.

Lyme disease

Lyme disease, caused by *Borrelia burgdorferi*, is increasingly reported from the USA, UK and Europe. It causes chronic illness with many similarities to those caused by the spirochaete of syphilis, *Treponema pallidum*, although its transmission is by a totally different route. At first, the patient experiences some fever and a characteristic rash in the form of a red margin spreading slowly out from the site of the tick bite (erythema migrans). The spirochaete then invades the body generally and may cause meningoencephalitis, facial paralysis, or spinal cord dysfunction similar to that seen in multiple sclerosis. Involvement of the heart muscle may occur. Also arthritis, at first acutely in the knee but later in a generalized pattern similar to that of rheumatoid arthritis, is particularly common in cases seen in the USA.

Diagnosis is made by serological methods, and treatment is with penicillin, tetracycline or cephalosporins.

Soft ticks

Soft ticks are blood-sucking ectoparasites of birds, bats and mammals including humans. They are found worldwide, particularly in the more arid parts of the tropics and subtropics, in the nest, burrow or habitat of the host. Soft ticks lack the hard dorsal shield found in the hard ticks. In contrast to the hard tick, the shape is round or oval and the

Fig. 14.19 Soft ticks, *Ornithodorus moubata*.

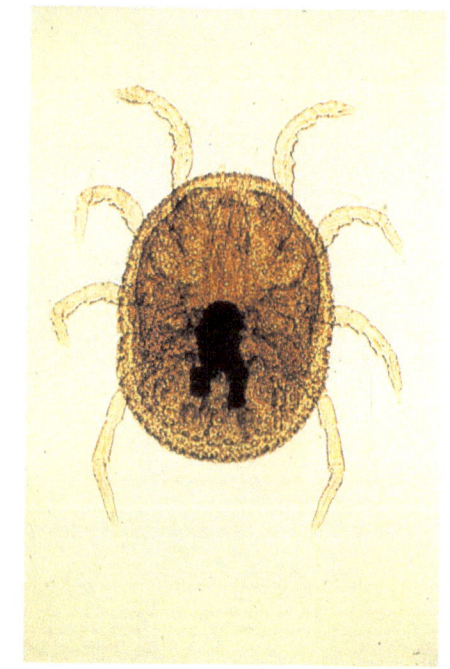

Fig. 14.20 Soft tick.

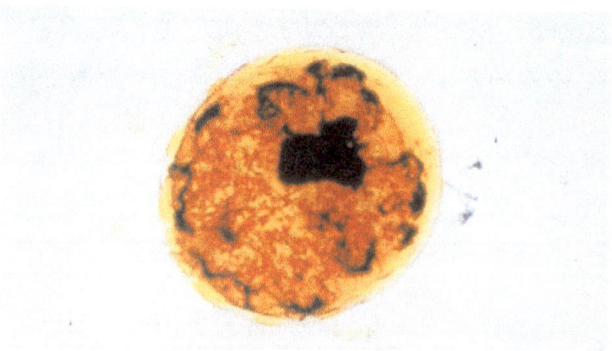

Fig. 14.21 Larva of soft tick.

integument is soft and leathery with a shrivelled appearance (Fig. 14.19), which becomes taut when full of blood. The mouthparts do not protrude and cannot be seen from above (Fig. 14.20).

Life cycle

The gravid female will lay several batches of eggs during her lifetime (each numbering ten to several hundred), in the habitat of the host. The six-legged larva that hatches is almost round in shape (Fig. 14.21) and will take a short feed when the host is in its habitat, but will typically remove itself before the host leaves. The life cycle is completed in the habitat of the host, moulting through several nymphal stages and each stage taking a blood meal. Species that commonly feed on humans, particularly within the *Ornithodorus moubata* complex in Africa, may be found around human habitation and may act as vectors of relapsing fever (Fig. 14.22). The larval stage of ticks within this complex remains in the eggshell after hatching and does not feed until it has moulted to the first stage nymph. There may be a total of four to seven nymphal stages; development from egg to adult may take about 3 months under optimal conditions, but all stages can resist starvation for long periods of 1 year or more.

Medical significance

The bite of a soft tick may be painful and the site may become ulcerated. While feeding, the tick will secrete a clear fluid from glands opening between the first and second pair of legs (coxal fluid), which will flood the space beneath the tick. If it has previously fed on an infected host, the coxal fluid will contain the spirochaetes of *Borrelia duttoni*, the causative organism of relapsing fever, which will enter the new host via the bite puncture. Once the tick has become infected it will remain so for life, and thus act as a reservoir of the disease as well as the vector.

Relapsing fever is characterized by the abrupt onset of high fever with severe headache, painful eye movements, muscle pains, cough, shortness of breath, jaundice and skin haemorrhages. Typically, the illness lasts for approximately 1 week, followed by a remission of a few days and then a relapse, which is less severe. Further relapses may occur at intervals. Tick-borne relapsing fever is usually less severe than the louse-borne disease caused by *Borrelia recurrentis*, but it may be neurotropic, causing facial paralysis, eye muscle paralysis, deafness, meningitis or encephalitis. Diagnosis is made by the demonstration of *Borrelia* in stained blood films. Treatment is effective with penicillin or tetracycline.

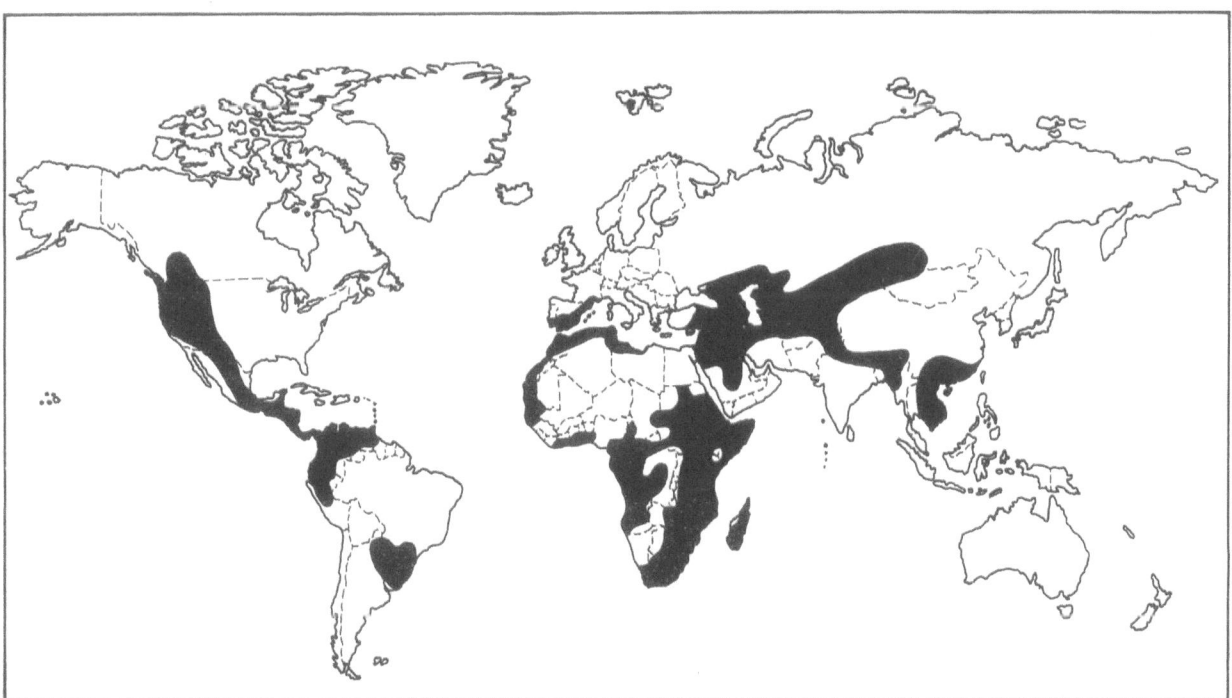

Fig. 14.22   Distribution of tick-borne relapsing fever.

15. Mites

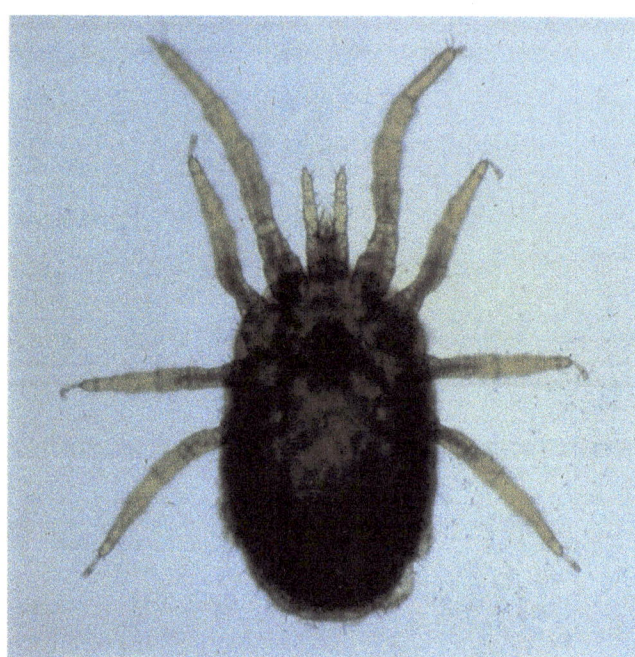

Fig. 15.1   Flour mite, *Acarus siro*.

INTRODUCTION

Mites are, in general terms, similar to ticks in appearance. They have a sac-like body, mouthparts grouped around the mouth opening as a capitulum, and eight legs in the adult stage. They are, however, with very few exceptions, much smaller in size, and their feeding habits and biology are very varied. Those of medical significance include the forage mite and the house-dust mite, both of which may cause allergies, the scabies mite causing the condition scabies by its burrowing habits, and the *Trombicula* mite, which is the vector of scrub typhus.

Several species of mite in the family Tyroglyphidae are pests of stored products and may cause an allergic response when humans come in contact with their faeces, dried bodies and cast skins. These include the flour mite *Acarus siro* (= *Tyroglyphus farinae*) (Fig. 15.1), and the furniture mite *Glycyphagus domesticus* (Fig. 15.2). Dermatitis may occur when storekeepers, grocers or bakers handle infested commodities, and it is even seen in office workers handling infested paper. A particular form of asthmatic allergy may occur in sensitized patients on inhalation of mite debris (Fig. 15.3) that is commonly found in bedding and furnishings, where the mite feeds on moulds that have formed on discarded flakes of human skin.

Occasionally, mites in the genus *Cheyletiella* (Fig.

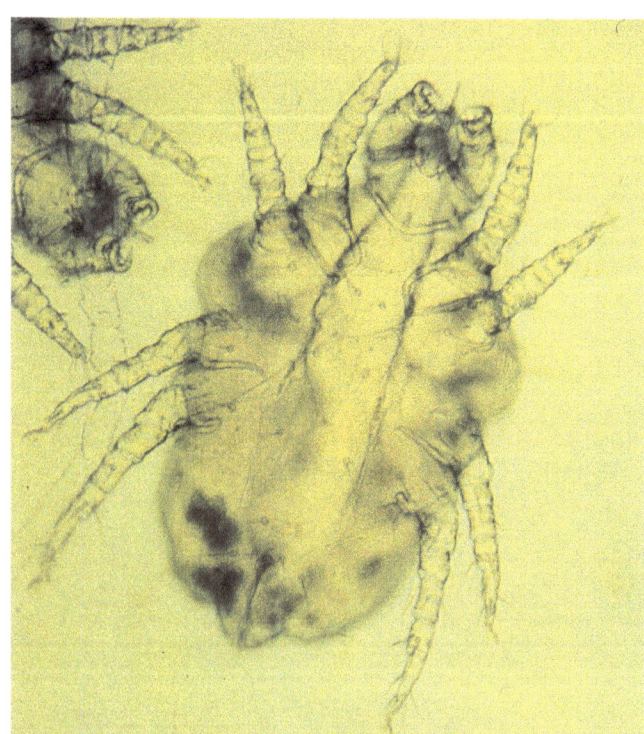

Fig. 15.2   Furniture mite, *Glycyphagus domesticus*.

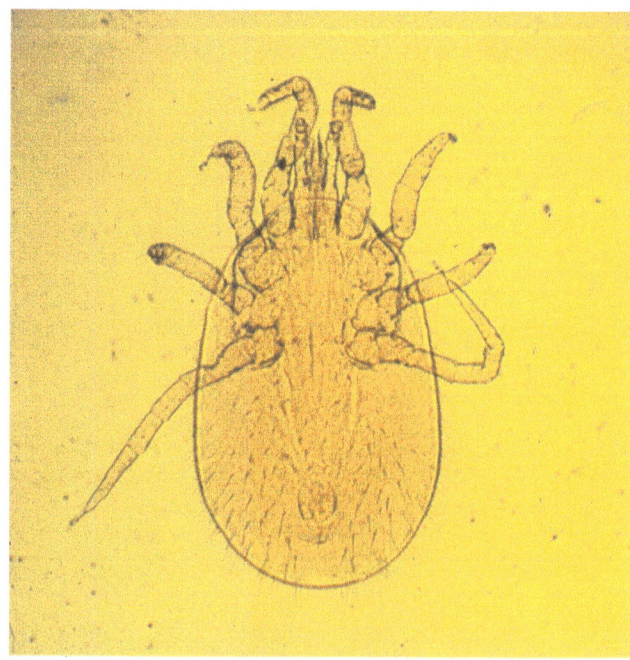

Fig. 15.3   House-dust mite, *Dermatophagoides pteronyssinus*.

15.4), which infest domestic pets, may cause an allergic response in humans after taking a blood meal. The presence of the mite can often be confirmed by microscopic examination of animal brushings. The hair follicle mite *Demodex folliculorum* (Fig. 15.5) inhabits human hair follicles, particularly of the eyebrows and face, but is rarely problematic.

HUMAN SCABIES MITE

The human scabies mite *Sarcoptes scabiei var hominis* is a permanent parasite in the subcutaneous tissue of humans, causing the condition scabies, which is seen throughout most parts of the world. Other varieties (which are not interchangeable) infest the skin of a wide range of domestic and wild animals, causing mange.

Description and life cycle

The human scabies mite (Fig. 15.6) is minute (less than 0.5 mm long) and disc-like with well-developed chewing mouthparts and eight short stumpy legs in the nymph and adult, terminating in suckers or bristles. The mite is yellowish-white in colour and bears several short spines on the dorsal surface. A scabies infestation will typically begin when a gravid female mite is acquired from an infested person, usually during skin contact. Early research indicated that mites could not be taken up from infested clothes or bedding since the mites cannot survive away from the host. The gravid mite (Fig. 15.7) burrows just below the surface and forms a tortuous burrow subcutaneously, defaecating and laying eggs at regular intervals. The presence of these foreign bodies will cause the allergic response known as scabies, which may be considerably more widespread than the mite infestation (Fig. 15.8). The eggs hatch into six-legged larvae, which form their own

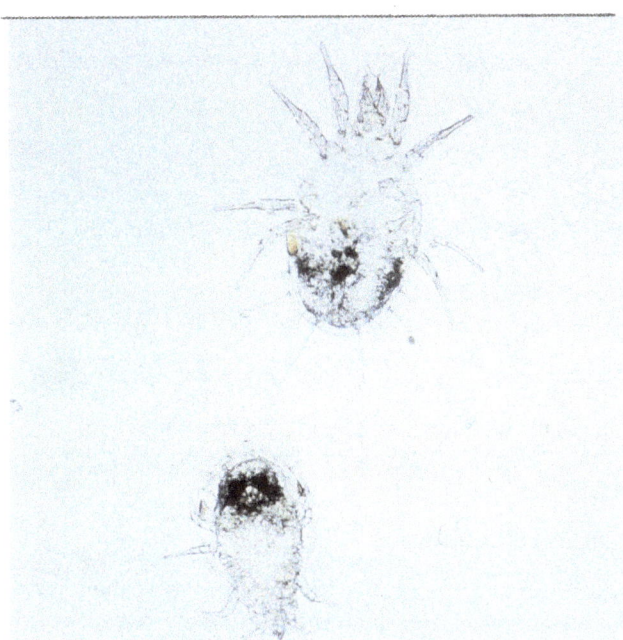

Fig. 15.4 Blood-sucking *Cheyletiella* species.

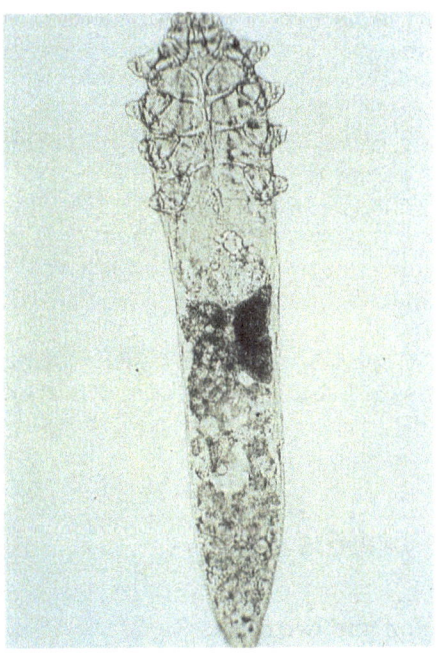

Fig. 15.5   Hair follicle mite, *Demodex folliculorum*

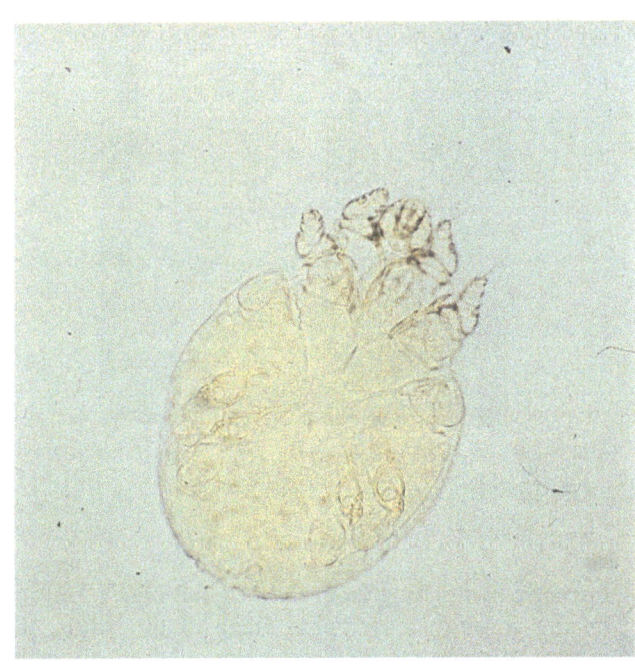

Fig. 15.6   *Sarcoptes scabiei*, the human scabies mite.

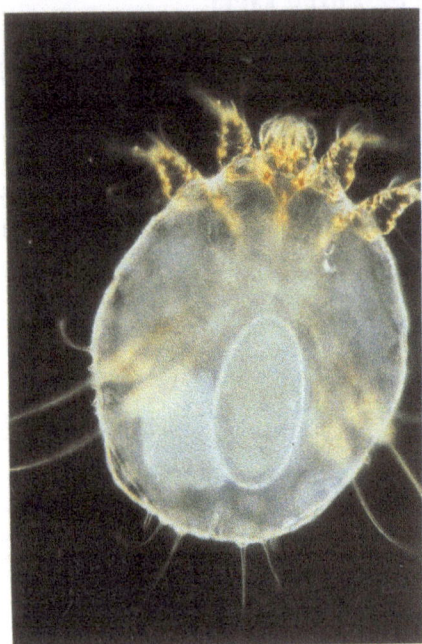

Fig. 15.7   Gravid female scabies mite.

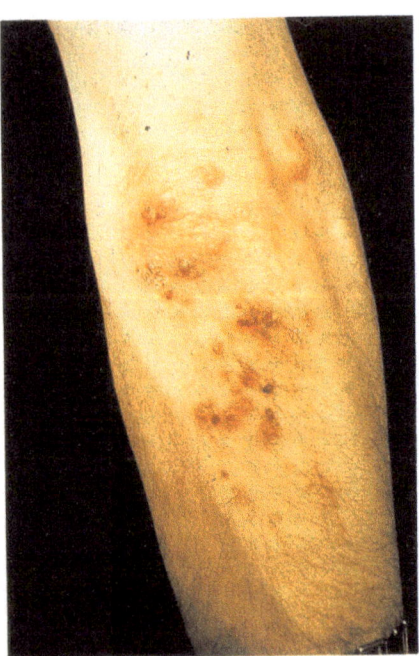

Fig. 15.8   Scabies, the allergic response to the human scabies mite.

surface. It is possible to tease a mite from a burrow with the use of a sterile needle. Dry skin that is scraped gently with a scalpel blade from an infested area may also reveal mites or their eggs (Fig. 15.9).

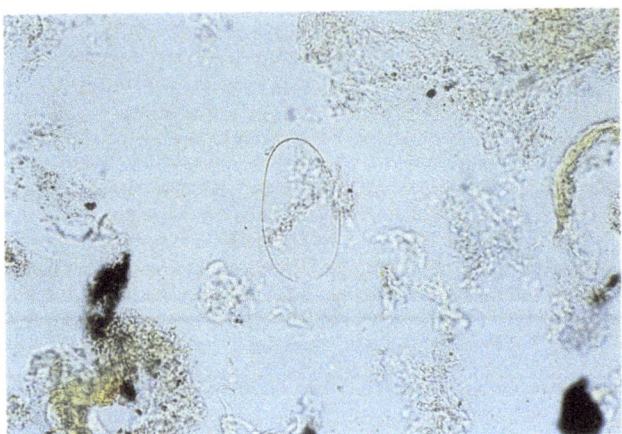

Fig. 15.9   Scabies mite egg in skin scraping.

Medical significance

Infestation of the skin with the human scabies mite *Sarcoptes scabiei* causes an intensely itchy papular and vesicular rash in reaction to burrowing adult females and their faeces. The rash is most commonly found on the hands (Fig. 15.10a), wrists, axillae, buttocks, groin (Fig.15.10b) and scrotum (Fig.15.10c); however, in Norwegian scabies it becomes generalized and severe, with crusting, affecting the face (Fig. 15.10d), particularly in patients with AIDS.

Scabies occurs worldwide and is transmitted by intimate personal contact. Diagnosis is made by demonstration of the mites in skin scrapings. Treatment is with topical 10% crotomiton or 10% Lindane, which are applied to all affected skin areas, and should be given to all close contacts of the patient.

In the tropics, scabies infestation may be complicated by skin infection with haemolytic *Streptococcus* spp., which can cause acute and chronic glomerulonephritis.

SCRUB TYPHUS MITE

Description and life cycle

Within the family Trombiculidae are several hundred species of mite worldwide, a few of which will

burrows, branching from the original one. Moulting occurs through a nymphal stage to the adult, the mite feeding by tunnelling continuously through the subcutaneous layers. It is thought that mating occurs on the skin when the adult mites burrow to the

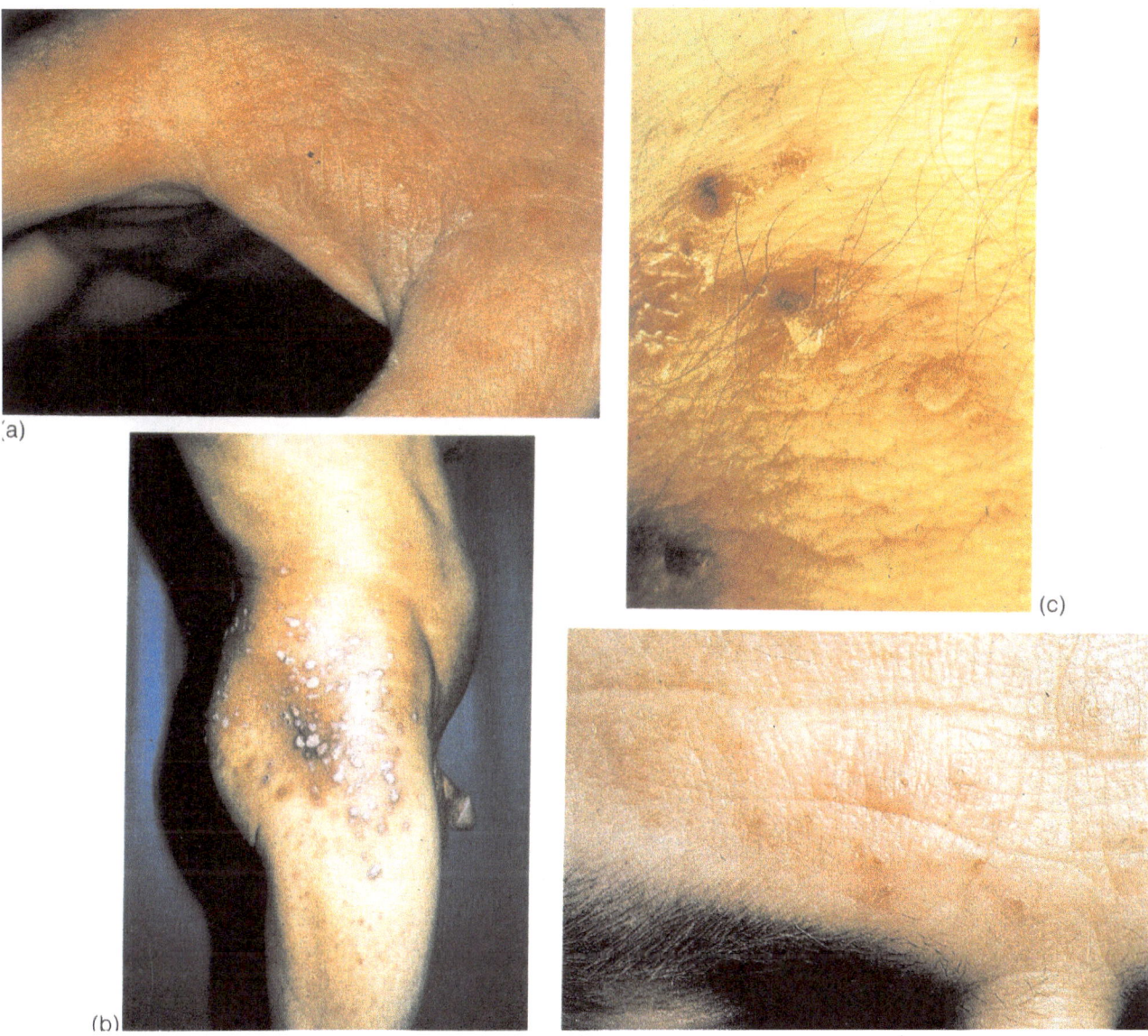

Fig. 15.10 Human scabies: (a) on the hand; (b) in the buttock; (c) on the scrotum; (d) on the face.

attack humans causing considerable irritation by the feeding habits of the larva. These mites, in the genus *Leptotrombidium* (= *Trombicula*), are free-living at all stages except the larval stage. Adults and nymphs live in the moist growth layers of vegetation and are sap-feeders. Eggs are laid in this environment but the hatched six-legged larva requires tissue from rodents or other animals.

The larva (Fig. 15.11) is minute (about 0.25 mm long), pale in colour, with hairs and bristles on the dorsal surface and trifurcate claws on each of the six legs. It will climb vegetation and cling to a passing host in much the same way as a hard tick. The mite crawls to a soft part of the host's body, particularly the ear, groin and axilla, and attaches itself to the tissue by means of the well-developed mouthparts. These secrete a histiolytic enzyme, which dissolves the tissue and which the mite subsequently imbibes. More fluid is continously secreted until a tube of tissue has dissolved, on which the mite has fed. This is known as a histiosiphon (Fig. 15.12). When the larva is fully fed it leaves the host, spending the rest of its life in a non-parasitic manner. These feeding habits may cause considerable irritation to the host. In temperate parts of the world the larva typically feeds in the autumn, giving rise to harvest itch or autumn itch, and is caused by species such as *Leptotrombidium autumnalis* in western Europe. The

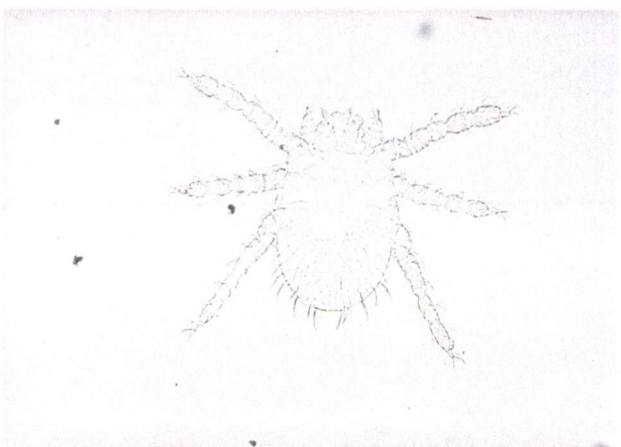

Fig. 15.11   The larval stage of *Leptotrombidium* spp., the scrub typhus mite.

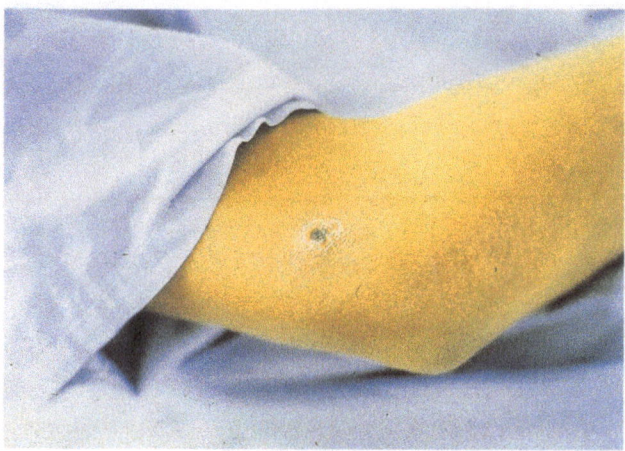

Fig. 15.13   Typical eschar.

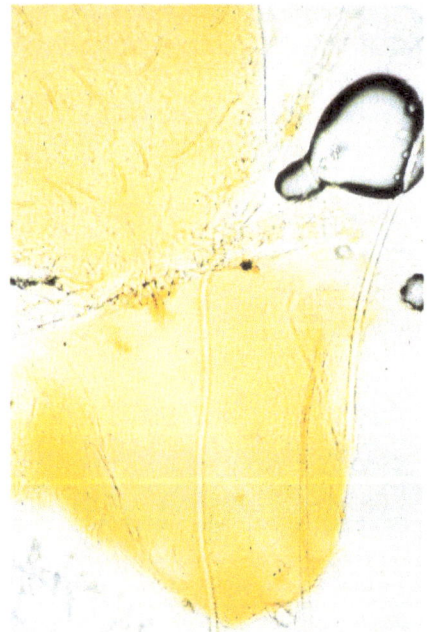

Fig. 15.12   Section through a larval mite feeding on tissue. The mouthparts and histiosiphon can be seen.

(usually a rodent), and the organism is taken up in the tissue on which the mite feeds. Thereafter the mite itself remains infected, passing the organism to the eggs (transovarial transmission) and subsequently through many generations. This is essential to the survival of the disease organism, since the vector is only parasitic in the larval stage and the larva feeds only once during its life cycle. Thus the mite colony itself will act both as vector and reservoir of the disease.

When a subsequent generation of larva feeds, perhaps on a human host, the organism will be passed on in the histiolytic fluid injected into the host. Mite islands, with the focus of scrub typhus infection, often occur in an area of nomadic cultivation that has been allowed to revert to secondary jungle. Rodents (the normal larval host) and mites are commonly found in this particular habitat (Fig. 15.15). Mites will also often harbour in piles of discarded palm leaves found in tropical plantations.

site of the bite often presents as a small, dark, necrosed area known as an eschar (Fig. 15.13).

Medical significance

In parts of India, China and southeast Asia (Fig. 15.14) species of *Leptotrombidium* act as vectors of scrub typhus, known also as mite typhus, Japanese river fever or tsutsugamushi fever. This is caused by the organism *Rickettsia tsutsugamushi*. A mite colony in a confined area (mite island) initially becomes infected when the larva feeds on an infected host

Clinical aspects

Scrub typhus is a febrile illness with concomitant severe headache, mental changes, macular rash (Fig.15.16), generalized lymph gland enlargement, splenomegaly and, in many cases, a small ulcer covered by a black crust ('eschar') at the site of a single bite of a larval mite (see Fig.15.13).

The diagnosis is often made in those exposed to the known habitat of the mite, whereupon patients can be serologically confined. Treatment is with doxycycline. People at high risk of infection because of their occupation may take doxycycline weekly as a prophylactic measure.

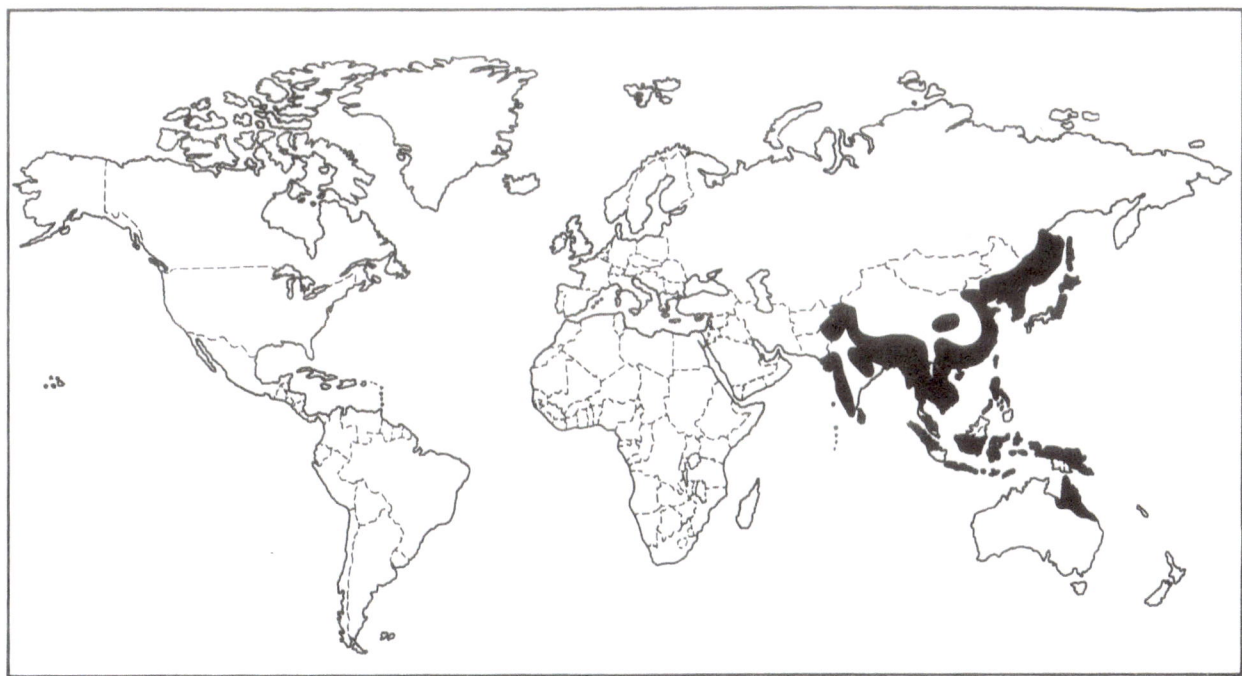

Fig. 15.14   Distribution of scrub typhus.

Fig. 15.15   Typical mite island in secondary jungle.

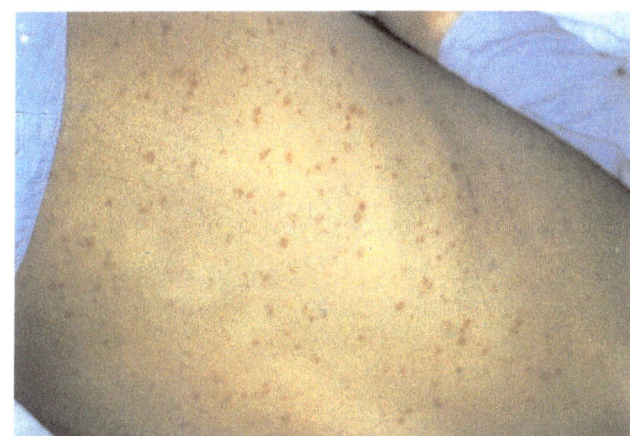

Fig. 15.16   Clinical appearance of scrub typhus rash.

16. Bites, stings and other forms of attack

In addition to transmitting disease and causing trauma by blood-sucking habits and other associations with humans, insects and other arthropods may have a wide range of effects on human health and well-being. This chapter discusses a miscellany of these.

ENTOMOPHOBIA AND DELUSORY PARASITOSIS

The appearance and presence of insects and other related arthropods arouses interest, curiosity and suspicion in most humans (Fig. 16.1). In some people, this may become an inordinate but very real fear, developing into a phobia (entomophobia) re-

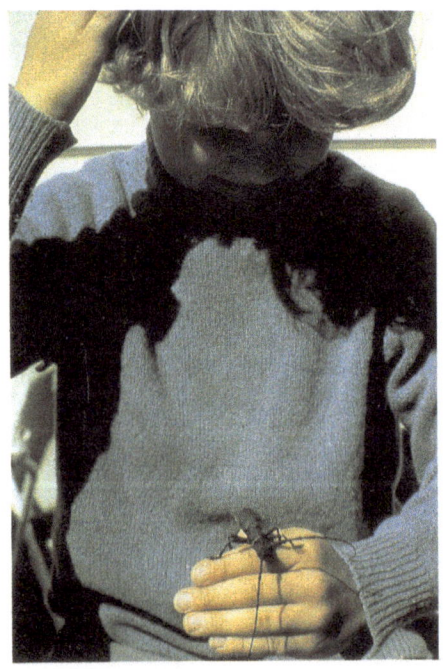

Fig. 16.1 Entomophobia.

quiring medical assistance. The phobia may be due to the fear of being bitten or stung, but more commonly it is because of the 'creepy-crawly' appearance of the creature, the long legs and antennae and its 'hairiness', a condition that would probably be increased considerably if human eyesight were more acute. Delusory parasitosis occurs when the patient is convinced that he or she is being attacked or infested by an insect or other arthropod. 'Bites' from some other cause may be present. Careful investigation may show no evidence of insect involvement. Sometimes the mental attitude is initiated by the chance finding of a single insect, perhaps a bedbug that has been brought accidentally into the patient's house, whereupon the patient becomes convinced of a much more widespread problem of which there is no evidence. The 'bites' may of course be an allergic response to a mite, a psocid, a thrips, or some other creature hardly visible to the naked eye, and a very thorough investigation of the environment, preferably by an entomologist, is essential. The patient must be asked a wide range of pertinent questions to ascertain whether any arthropod contact has been possible or likely, and brushings from surfaces and fabrics, and from domestic animals in the environment, must be carefully examined under the microscope for evidence of the presence of arthropods. The basis of both the phobia and the delusion is often a fear of the unknown and, although the patients may not respond to a reasonable argument, an identification of the cause is often helpful.

CABLE BUG OR CABLE MITE

With the considerable increase in the use of electronic equipment and synthetic fibres, a phenomenon is emerging, which has been termed 'cable bug' or 'cable mite'. The patient has apparently been bitten, and shows ample evidence of this, often scratching the 'bites' until they become secondarily infected (Fig. 16.2).

An exhaustive investigation may reveal no evidence of arthropod involvement, but debris examined under the microscope may include spicules of fibreglass and synthetic fibres. The condition typically occurs in premises where there is a high level of static electricity (e.g. in computer rooms, with floor coverings, insulation or filtering equipment containing synthetic fibres). This charge attracts spicules to exposed areas of the skin where they set up an allergic reaction. A spicule may actually pierce the skin in much the same way as an insect's proboscis, and may penetrate a stockinged

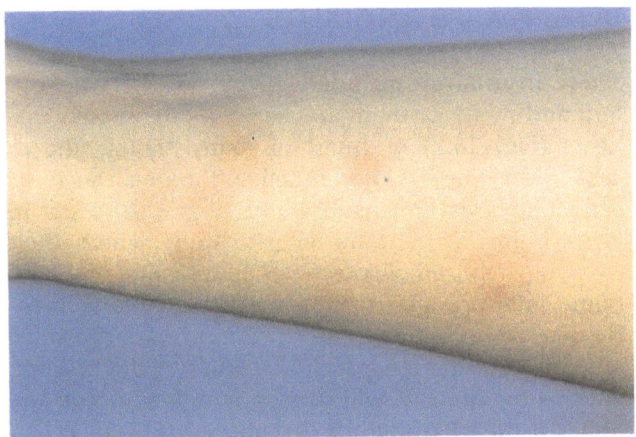

Fig. 16.2 'Bites' of 'cable bug' or 'cable mite'.

URTICARIA CAUSED BY CATERPILLARS

A simple allergic reaction may be caused by coming into contact with the hairs and spicules of many Lepidoptera, that is, butterflies and moths (Fig. 16.3). This can be seen throughout most parts of the world. In the Mediterranean, the processionary caterpillar (larva of the moth *Thaumatopoea wilkinsoni* and other species) spins a web in the vegetation in which it lives (Fig. 16.4). If the web, with its occupants and their cast skins, comes into contact with human skin, an allergic rash may result. The larvae are named processionary caterpillars because they process down the trunk of the tree when ready to pupate, in order to undergo further development at ground level (Fig. 16.5). The puss moth of western Europe (Fig. 16.6) may cause a similar urticaria.

leg. A simple remedy may be to reduce the static electricity by humidifying the environment, although care must be taken not to dampen it to the extent of allowing moulds to grow on discarded skin flakes, encouraging house-dust mites and an alternative allergy.

(a)

(b)

Fig. 16.3 **Lepidoptera** (a) Larva of *Philudoria potatoria*. (b) Larva of *Sibine sp.* from Belize.

Fig. 16.4 **Processionary caterpillar.** (a) Web of the caterpillar (Cyprus).(b) Nest with cast larval skins.

Fig. 16.5 Mature *Thaumatopoea wilkinsoni* caterpillars processing before pupating.

Fig. 16.6 (a) The Oak Eggar larva, *Lasiocampa gueraus.* (b) Web of Lackey moth, *Malacosoma neustria* in vegetation.

BEES, WASPS AND ANTS

These insects are included in the order Hymenoptera and many of them are social, living in colonies, nests (Fig. 16.7) or hives of many thousands, in castes serving various functions. Some, such as Pharaoh's ants and garden ants (Fig. 16.8) occur in large numbers in domestic situations where their foraging habits may pose a public health problem. Some ants are able to bite with their strong mandibulate mouthparts (Fig. 16.9) and can cause an allergic urticaria (Fig.16.10). Certain wild species, such as the fire ant, will shelter in hollow thorns and attack if their harbourage is damaged (Fig. 16.11).

The ovipositor of many of the Hymenoptera has become adapted to form a poison sting that is used to inject venom into prey or can be used as a defensive mechanism. Reaction to these stings in humans is very varied depending on the level of

Fig. 16.7 Wasp nest.

Fig. 16.8 Garden ants swarming.

Fig. 16.9 Wood ant showing strength of the mouthparts.

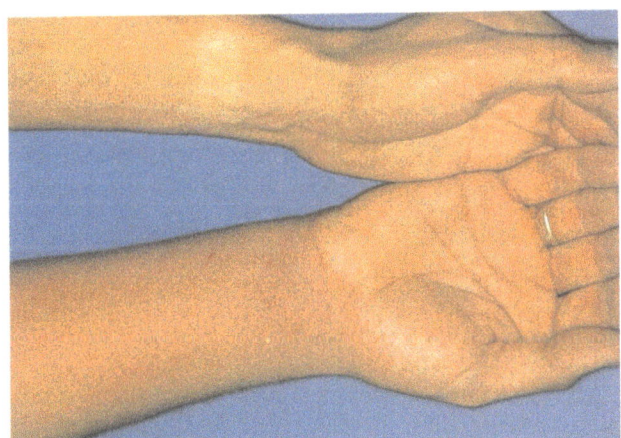

Fig. 16.10 Bites from ants *Formica sanguinea*.

Fig. 16.11 Thorns harbouring fire ants within (Belize).

sensitivity, and while it is usually mild, occasional cases of hypersensitivity can occur, resulting in death. Bees (Fig. 16.12) will leave their sting in the site of the attack, whereas wasps and hornets (Fig. 16.13) will remove it. Several hover fly species (16.14) can mimic bees and wasps but these are completely harmless and only have one pair of wings (Diptera) compared with the two pairs of the bees and wasps.

Fig. 16.12 *Apis mellifera*, the honey bee.

Fig. 16.13 The hornet, *Vespa crabro*.

Fig. 16.14   The hover fly, *Eristalis tenax*.

exist, ranging from 1.5 to 20 cm in length (including the telson). The chelicerae of the scorpion are used to hold its prey while injecting a paralysing fluid into it and feeding on the fluid content. Scorpions, in common with many arthropods, will not attack humans unless provoked. They are typically cryptic creatures, sheltering under stones, tree trunks and other such places, but may be encountered by accident in domestic situations into which they may have inadvertently strayed (e.g. in clothing, bedding and footwear).

The colour of each scorpion species will correlate with its environment, those of desert regions being sandy-coloured (Fig. 16.16), whereas those inhabiting forests will be darker (Fig. 16.17).

SCORPIONS

Scorpions (order Scorpiones, within the class Arachnida) are found in many parts of the tropics and warm temperate zones, in both humid and arid areas.

Adults have eight legs, large claws (chelicerae, which are the outer pair of mouthparts), and a flexible tail or telson, at the tip of which there is a needle-like sting (Fig. 16.15). Nearly 400 species

Fig. 16.16   The scorpion (Sudan).

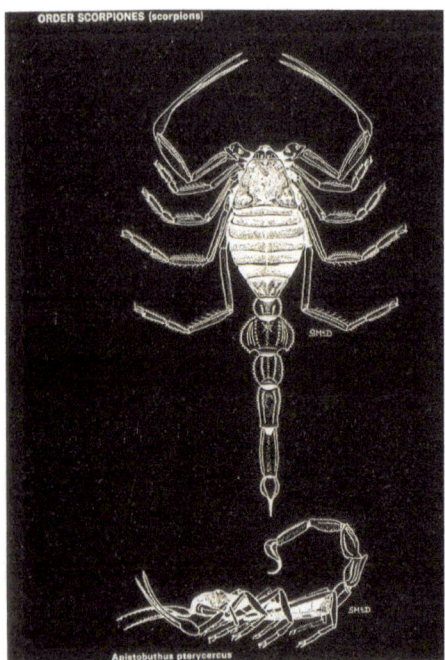

Fig. 16.15   Order Scorpiones.

Fig. 16.17   The scorpion (southern France).

SPIDERS

Spiders (order Araneae, class Arachnida) in the adult stage have the body divided into two regions (fore and hind), with paired mouthparts at the front end and eight legs attached to the forebody. There are several thousand species worldwide and the size range is considerable, the smallest being no more than a few millimetres in length and the largest having a leg span of up to 20 cm. Spiders capture their prey using the mouthparts, sometimes causing paralysis by injecting venom. Fluid from the captured prey is then extracted by sucking. The venom of a few species of spider is particularly toxic to humans. The large hairy tarantula-type of spider (Fig. 16.18) that is found in many tropical and Mediterranean countries is rarely dangerous, whereas the bite of the smaller *Latrodectus* species (Fig. 16.19), found in most tropical and subtropical regions, can be particularly toxic and sometimes fatal. Other species such as the brown recluse spider *Loxosceles* can cause considerable necrosis at the site of the bite (Fig. 16.20).

Camel spiders (solpugids, 'sun spiders') are not classified as true spiders and are not venomous, although there have been alarming reports of their effect on humans. These spiders are carnivorous and their large jaws (Fig. 16.21) are often contaminated with pathogenic bacteria. In humans, a bite from this spider may result in extensive and dramatic secondary infection, whereupon the camel spider is credited with eating the tissue.

Fig. 16.19 Black widow spider, *Latrodectus mactans*.

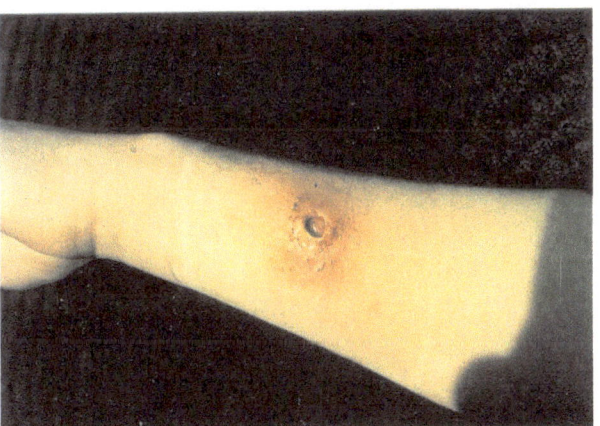

Fig. 16.20 Necrosis caused by bite of brown recluse spider, *Loxosceles reclusa* (USA).

Fig. 16.18 Tarantula (Belize).

Fig. 16.21 Camel spider.

129

CENTIPEDES

The most typical arthropod is perhaps the centipede, of which there are several hundred species worldwide, making up the order Chilopoda. Centipedes are elongate and flattened dorsoventrally, with segments ranging from 15 to over 100 in number, each with one pair of legs (Fig. 16.22). The length of the centipede varies from a few millimetres to over 25 cm. The front pair of legs is adapted to form fangs attached to venom glands, which the centipede uses to inject poison into its prey. A few of the larger species, such as *Scolopendra morsitans* (Fig. 16.23), may inflict an unpleasant bite to humans.

Millipedes, in the order Diplopoda, can be easily differentiated from centipedes by the circular cross-section of the body and the presence of two pairs of legs to each apparent segment. Millipedes are vegetarian, without venom glands. Although they sometimes attain lengths of up to 25 cm and occur in large numbers (Fig. 16.24), they are of little medical significance. One or two species are able to secrete a vesicant fluid from the glands on the dorsolateral surface, which may cause irritation and blistering.

Some beetles, for example *Paederus* species, as well as vesicant plants and the nematocysts of jelly-fish (Fig. 16.25), may cause a similar effect.

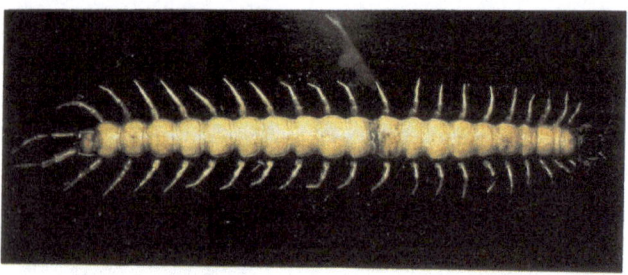

Fig. 16.22   Centipede.

Fig. 16.24   Mass of millipedes (Nepal).

Fig. 16.23   Centipede, *Scolopendra morsitans*.

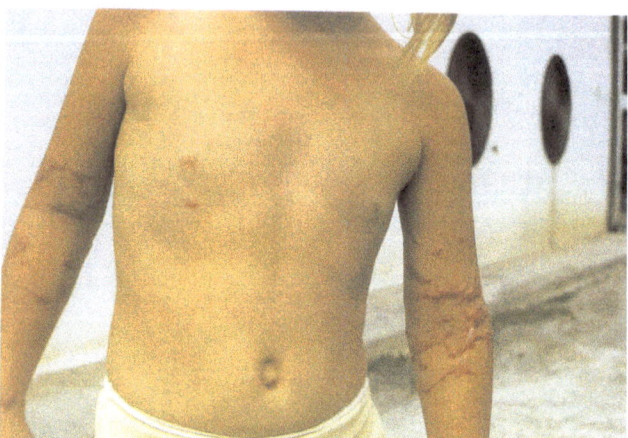

Fig. 16.25   Jellyfish stings.

17. *Control of medically significant arthropods*

The detailed and technical aspects of vector control are beyond the scope of this work. However, the basic approach to the subject is so fundamentally important that no book on medical entomology would be complete without some consideration of the principles of control. The reduction and eventual eradication of arthropod-borne disease depends on two factors: the first is the elimination of the disease organism from its human host or reservoir by protection or treatment; the second, and perhaps more important, is an attack on the arthropod vector, for without the vector there can be no disease.

PERSONAL PROTECTION

The disease organism can be killed or rendered non-infective in the patient by the use of therapeutic drugs. This will at least ensure that the organism is not available to the vector to transfer to a new host, although the patient will already have suffered symptoms of the disease that may be irreversibly incapacitating or even fatal.

Development of the organism can be arrested as soon as it enters the patient by the use of prophylactic drugs such as proguanil and chloroquine, or vaccination. This ensures to a greater or lesser extent that the patient will suffer no ill-effects from the disease, and also that the organism does not undergo further essential development and is thus not available in an infective form to the vector.

It is clearly more satisfactory if the patient can avoid being bitten by the arthropod, since this prevents the pain and irritation of the bite as well as the risk of disease. The use of insect repellents, particularly those containing diethyltoluomide (DEET), is effective if all exposed skin is treated. A formulation containing less than 50% DEET is recommended to avoid any risk of side-effects, and long duration (micro-encapsulated) preparations are now available, giving protection for up to 8 hours. The area of exposed skin should be reduced to a minimum when insects are biting. For example, arms and legs should be covered at dusk, during the night and at dawn to avoid bites from mosquitoes. However, these insects will bite through thin layers of clothing, and other arthropods such as hard ticks and scrub-typhus mites crawl over outer garments to find access to the skin.

Treatment of clothing, particularly socks and trousers, with a miticide such as dibutyl phthalate (DBP) or an insecticide such as permethrin is recommended. As with all protective measures, the manufacturer's instructions on the container label must be read and followed. So-called electronic (ultrasonic) insect repellers, which give out an audible high-pitched whine, have been shown to be completely ineffective.

It often appears that some people are bitten more frequently than others, but the factor that attracts blood-sucking insects has not been shown conclusively. It is probably a combination of factors such as movement, body temperature, carbon dioxide exhalation and colour and texture of skin, as well as chemicals given off by the skin and body surfaces. Certainly the state of immunity or sensitivity to the injected anticoagulant saliva or other proteinaceous material will govern the host's reaction. In general, fair-skinned people, the young and the elderly may be more highly sensitized, whereas darker skinned people may have acquired an immunity. Nevertheless, the response can be extremely varied.

Sleeping under a mosquito net at night is a very effective way of avoiding bites and hence vector-borne disease. The mesh size of the net, however, may allow smaller blood-sucking flies such as sand-flies to enter, although once fully engorged with blood they are unable to get away. Impregnation of the bednet with permethrin is an effective way of avoiding this and it has been shown conclusively that even damaged nets, provided they have been impregnated, are effective against mosquitoes. Impregnation of curtains and other fabrics is also advocated. Where a mains supply of electricity is available, a small hot-plate (Fig. 17.1) may be used to vaporize a tablet containing a pyrethroid insecticide,

Fig. 17.1 Electrical insecticide vaporizer.

killing insects in the surrounding atmosphere. Mosquito coils serve a similar function.

VECTOR CONTROL

The most satisfactory method of arthropod-borne disease control is to combine methods directed towards treating and protecting the human host/reservoir with control measures directed against the vector. Any successful control programme should involve five stages:

1. Survey
2. Planning
3. Control measures
4. Monitoring
5. Follow-up.

Identifying the species

Before satisfactory measures can be undertaken it is essential to identify the species of vector involved so that control can be directed in the most effective way. For example with mosquito-borne disease, control organizers need to know which species are responsible, at what time of day and at what location they feed, whether they enter houses and rest inside before or after feeding, what type of environment provides breeding sites, and so on. Collection and identification of the species will provide much of this information already available in the literature. Various methods of collection for different arthropods can be seen in Fig. 17.2.

(a)

(b)

c)

Fig. 17.2 Collection and identification of vectors. (a) Adult mosquitoes are collected with an aspirator or pooter. (b) A USA army light trap, which can be baited for increased effectiveness with solid carbon dioxide attracts flying insects. (c and d) Mosquito larvae are scooped from their breeding sites with a soup ladle. (e) The mosquito larvae can be bred through to the adult for easier identification. (f) Specimens for examination for parasites or virus isolation are sorted on a cold table. (g) Cockroaches and other crawling insects can be collected in petri dishes or (h) in sticky traps that are left in position overnight and examined in the morning. (**Continued over**)

(d)

(e)

(g)

(f)

(h)

Fig. 17.2 continued.

Physical measures of control

Once the incriminated species has been identified and the relevant information assessed, the control programme can be planned. When an arthropod attacks humans or becomes a public health pest, it is seeking one or more of the essential requirements of food, water, shelter and warmth. If it can be deprived of any of those necessities, control will be rendered easier and more effective. Thus purely physical measures may play an important role in any control programme. For example, aquatic breeding sites for mosquitoes should be drained or channelled (Fig. 17.3) so that the water flow is too swift to allow egg-laying. Damaged or leaking pipes (Fig. 17.4) may provide water for mosquito breeding sites, or provide essential water for cockroaches in a domestic situation.

Harbourages can be removed or reduced by better design of food preparation areas (Fig. 17.5) or the use of more efficient building materials. For example, atap roofing, which provides ideal hiding places for cone-nosed bugs, can be replaced with corru-gated iron sheets, and walls can be plastered to reduce cracks and crevices (Fig. 17.6).

Basic standards of hygiene and good housekeeping are essential if a pest-free or vector-free environment is to be maintained. Accumulations of rubbish and detritus, animal and human faeces, carcasses and other decomposing organic matter (Fig. 17.7) will attract flies and cockroaches as feeding and breeding sites, and will also act as a source of potential pathogens. These hazards should be removed before insecticidal treatment is initiated. Electric fly killers (EFKs) (Fig. 17.8) are a useful supplement to good hygiene and insecticide treatment. Flies are attracted to an ultraviolet (UV) light and electrocuted by a high voltage grid placed in front of the light source, the dead flies being collected in a tray at the base of the grid. Siting of the EFK away from competing sources of UV light (daylight and fluorescent lamps) is important.

Physical measures in isolation may not be sufficient to reduce the pest or vector problem to an acceptable level and the use of chemical or biological control agents may be required. The most effective

(a)

(b)

Fig. 17.3 Prevention of mosquito breeding sites. (a)
Blocked drain provides mosquito breeding site. (b) Well-
channelled drain.

Fig. 17.4 Damaged irrigation and drainage pipes.
Proper maintenance is important as they provide breeding
sites for (a) mosquitoes or (b) water supplies for
cockroaches.

Fig. 17.6b   Walls can be plastered to reduce harbourages.

Fig. 17.5   Good hygiene and practicable design of catering facilities. This reduces cockroach harbourages and makes good housekeeping easier.

Fig. 17.7   Decomposing organic matter. Any matter of this sort will attract flies.

Fig. 17.6a   Selection of building materials. Replacement of roofs with corrugated iron.

Fig. 17.8   Electric fly killer. When correctly sited and maintained this is a useful contribution to control.

agent and the optimal method of application should be considered carefully, to ensure that the insect and the insecticide interact, resulting in effective control.

Insecticide control

Chemical insecticides are designed to kill insects in a variety of ways. They may act as a stomach poison when formulated as a bait that is eaten, but more commonly they will affect the nervous system of the insect, passing through the spiracles, feet or cuticle. The impervious nature of the cuticle may be reduced by incorporation of a desiccating agent in the formulation. The insecticide may be formulated for application as an aerosol mist or fog through which the insect will fly (Fig. 17.9) when its action may be immediate (knockdown) but with no long-lasting

residues in the environment and food chains, resulting in unacceptable toxic effects in non-target species.

The further development of insecticide groups, through the organophosphates (e.g. malathion and fenitrothion), carbamates (e.g. bendiocarb, propoxur) to the pyrethroids (e.g. permethrin, cyper-

(a)

Fig. 17.9 A fog of knockdown insecticide. On flying through this the insect will be killed.

effect, or it may be designed for application to surfaces or materials on which the insect will settle or crawl, providing a persistent or residual killing action over several days or weeks. The earliest insecticides were essentially inorganic chemicals that acted as stomach poisons, and were as toxic to mammals and other non-target species as to their intended target. More specific insecticides were plant derivatives such as pyrethrum from species of chrysanthemum. These have a knockdown action that is enhanced by the addition of a synergist such as piperonyl butoxide. The organochloride insecticides such as DDT and BHC (Lindane) were developed over 50 years ago. These are very effective and economical to produce, but their long residual life, often of many months, causes an accumulation of

(b)

Fig. 17.10 Wettable (water dispersible) powders. (a) Internal application with a compression sprayer. (b) Wettable powders are also effective when applied externally to porous surfaces.

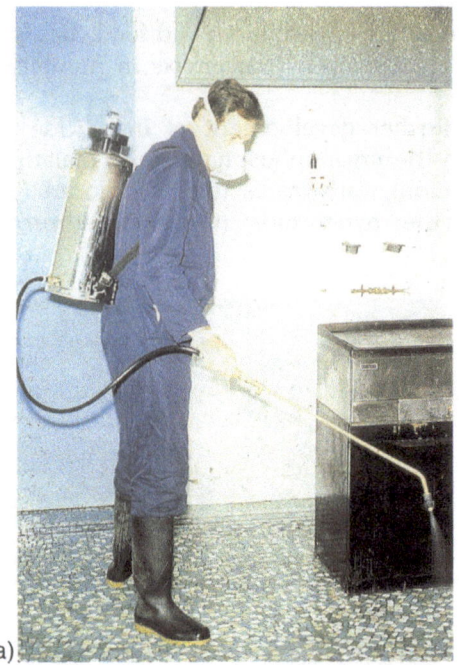

(a)

b)

Fig. 17.11 Control of insects. (a) Emulsion concentrate applied to non-absorbent surfaces with compression sprayer. (b) Insecticide applied as a mist or fog.

cavity treatment and sites such as electrical equipment where fluids are unacceptable (Fig. 17.12); powders for personal application; smokes for confined spaces; gels and lacquers; and products for impregnation of clothing, nets, curtains (Fig.17.13) and so on. Baits incorporating an insecticide (Fig. 17.14) are designed as stomach poisons and residual insecticides may also be applied as an aerosol to surfaces. Knockdown (non-persistent) insecticides are also formulated for application in aerosol form to the space through which insects fly, and for dispersal as a mist or fog.

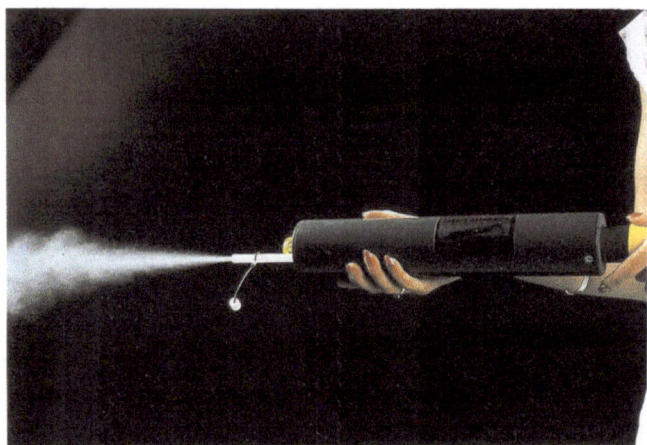

Fig. 17.12 Dust applied with a dust-gun.

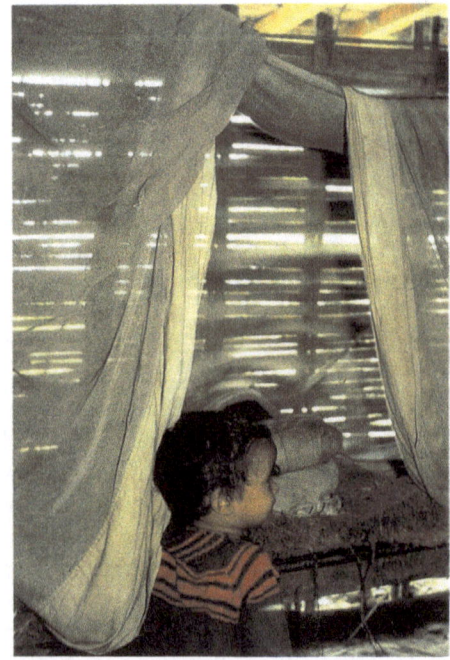

Fig. 17.13 Bed nets, curtains and clothing can be impregnated with permethrin.

methrin) was aimed at combining more specific toxicity to the target species with the lowest possible mammalian toxicity, and an accepted persistence and acceptable breakdown period to avoid excessive residues in the environment and in non-target organisms. The range of formulations is considerable, and includes emulsions and flowable concentrates for use on non-porous surfaces; and wettable (water-dispersable) powders for absorbent surfaces such as brickwork, plaster and fabrics, applied via a compression sprayer (Fig. 17.10). Emulsion or flowable concentrates are effective in less-absorbent surfaces, applied with compression sprayers (Fig. 17.11a) or as a fog or mist (Fig. 17.11b); dusts for

Fig. 17.14 Insecticidal bait applied against cockroaches.

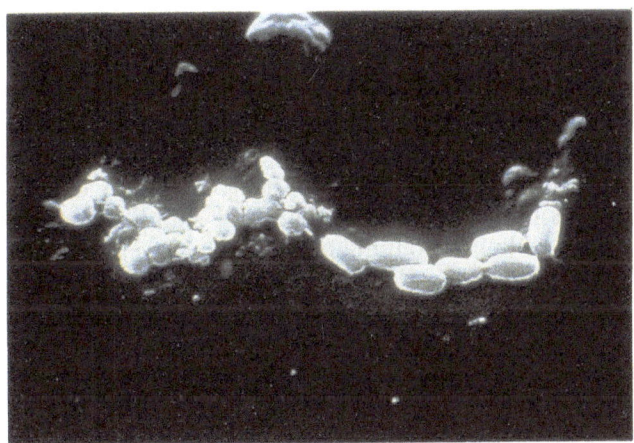

Fig. 17.15 *Bacillus thuringiensis israelensis* (Bti) crystals.

Insect growth regulators

The synthesis of insect growth regulators (IGRs), that are produced naturally by the insect to control development, enables much more specific control to be achieved. For example, application of methoprene to the water in which mosquito larvae are developing or to the harbourages of flea larvae will prevent the insect from reaching adulthood. Cockroaches treated with hydroprene will remain as nymphs and thus be unable to reproduce. The use of IGRs in control programmes can be very effective if correctly applied. Control is highly specific but may take longer to achieve than with conventional insecticides. However, for certain pests such as Pharaoh's ants, it may be the only effective method.

Insecticidal bacteria such as *Bacillus thuringiensis var israelensis* (Bti) (Fig. 17.15) and *Bacillus sphaericus*, which act as stomach poisons causing the gut to disintegrate, are also extremely specific and effective when taken up by the insect.

Mosquito breeding sites can be treated with insecticide or simply sprayed with a light oil (Fig. 17.16). A novel and very effective method (if somewhat ecologically unacceptable) is to soak parcels of hessian in diesel oil and throw them into water in which mosquitoes are breeding (Fig. 17.17). Mosquito larvae may also be controlled by adding larviporous fish (*Gambusia*) to the water. These are bred specifically for this purpose. Horseflies (clegs) may be collected on sticky boards (Fig.17.18).

Fig. 17.16 Mosquito breeding sites treated with insecticide or light oil.

Fig. 17.17 Oiled hessian parcels.

139

Fig. 17.18 Coloured boards to attract horseflies *Haematopota pluvialis.* The horseflies are subsequently trapped by non-drying adhesive.

Fig. 17.19 Cockroach egg-cases (oothecae) in corrugated cardboard.

General measures

It is important to record all use of pesticides and any control action taken in a pest or vector control book or file. Even in the field makeshift recording methods can be adopted, such as the dates of mosquito spraying noted on hut doors.

Danger of reinfestation is constant. Cardboard and other containers often carry cockroach egg-cases into previously uninfested premises (Fig. 17.19). It is important to treat all interconnecting ducting and shafts (Fig. 17.20) with residual insecticide.

Perhaps the most important aspects of any vector or pest control operation is effective communication and good public relations. The cooperation of the local population is essential and the significance and need for the programme must be explained.

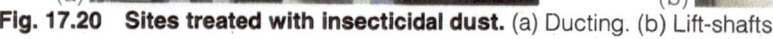

(a)

(b)

Fig. 17.20 Sites treated with insecticidal dust. (a) Ducting. (b) Lift-shafts.

Index

Numbers given in **bold** are figures, and numbers given in *italic* are tables.